SpringerWienNewYork

Zvonka Zupanič Slavec

NEW METHOD OF IDENTIFYING FAMILY RELATED SKULLS

Forensic Medicine, Anthropology, Epigenetics

Forewords by
Gertrude Hauser, Eugen Strouhal
and Igor Grdina

SpringerWienNewYork

Z. Zupanič Slavec, MD, PhD
Associate Professor of medical History
Institute for History of Medicine
Medical Faculty, Ljubljana University
Zaloška 7a
1000 Ljubljana
Slovenia

Printing was supported by
Ministry of Education, Science and Sport of the Republic of Slovenia

SpringerWienNewYork is a part of Springer Science+Business Media
springeronline.com

Design and typesetting: Metka Žerovnik
Printing: Grafika Soča d. d., Slovenia
Translation: Alan McConnell-Duff, MA
Photography: Viktor Berk

Printed on acid-free and chlorine-free bleached paper

SPIN: 11009894

Library of Congress Control Number: 2004107459

ISBN 3-211-22044-5 SpringerWienNewYork

To my family

Forewords

WITH HER MONOGRAPH Zvonka Zupanič Slavec impressively demonstrates by stepwise scientific analysis how identification of the 18 skulls attributed to the Counts of Celje has been achieved. It is already evident from the author's acknowledgements how difficult this task was. In 1811 the Minorite church burned, where 18 members of the noble family of the Counts of Celje had been buried between 1350 and 1456 when the last count died. As is so often the case, the disturbance and reburial of skeletal remains led to the loss of any written identification that may have indicated identity. Thus the fire initiated a succession of losses of information as to the identity of the 18 individuals.

With the exception of the first and the last count it was not known any more which family members had been buried in the original family vault. Studies of identification of historical skeletons make use of various research strategies drawn from a great variety of disciplines such as: physical anthropology, radiology, stomatology, anatomy, paleopathology, genetics, history and particularly so from forensic medicine. All these tools were applied by Zvonka Zupanič Slavec in establishing the age, sex, morphology, pathology and epigenetic similarity traits of the 18 individuals for whom this identity information was no longer available. These findings were combined with information from written records to produce a remarkably full description of the genealogical positions of the individuals.

Thus, notwithstanding the many difficulties in work of this kind Zvonka Zupanič Slavec achieved her goal. She concludes with a discussion of the usefulness of such studies for elucidating sociocultural aspects of human life in the late medieval time of this region of Middle Europe. This work, based on the latest epigenetic research, is also highly relevant for modern identification studies.

Prof. Gertrude Hauser, PhD
Histologic-Embryologic Institute
Medical Faculty, University Vienna, Austria

THE PRESENT BOOK was written by Professor of the History of Medicine at Ljubljana University, Dr. Zvonka Zupanič Slavec, renowned author of a number of important scientific contributions in that field. This time, however, she decided to identify historical persons in a group of 18 human skulls (except for one without mandibles) deposited in the Minorite Church of the Assumption of the Virgin Mary in Celje, Slovenia. After the fire and renovation in 1811, they were discovered in the crypt of the church. According to tradition and some written evidence they have been considered members of the important medieval families of the Counts of Celje.

A similar endeavour is usually performed by a multidisciplinary team of historians, physical anthropologists, radiologists, forensic doctors and palaeopathologists. In this case, however, the author was able to master the basic methods of all these branches and has most successfully accomplished the task.

In spite of the current feelings that today only molecular DNA analysis is an exact identification method, and, in case that DNA cannot be isolated, it would be better to give up the case, the author has also used for the same purpose older, classical methods. These, taken multidisciplinarily, can serve the same purpose as the reference method.

In her case especially, a recent method of study of epigenetic traits proved to be very useful for identification of the skulls in connection with historical and other data. This method can be applied in similar cases all over the world.

Not less important are the palaeopathological results, based on the author's knowledge of pathology, physical anthropology and medical history.

Part of the results are also useful for studies of the socio-cultural aspects of human life, giving back to the identified people their »faces« by application of the Gerasimov method of reconstruction of soft tissues of the skull.

The book brings new facts bearing on the history of the Counts of Celje, which are of the utmost importance for Slovene national history. This study will interest not only specialists of the above-mentioned scientific branches, but also a wide international circle of readers.

Prof. Eugen Strouhal, MD, PhD
Institute for the History of Medicine and Foreign Languages
First Medical Faculty, Charles University
Prague, Czech Republic

Author's Acknowledgements

This long continuing research, which, like a kaleidoscope with different viewing angles presenting each time a new visage, through its expressly interdisciplinary nature has drawn me into contact with numerous interesting experts and interlocutors. By their encouragement and responsiveness they greatly assisted me in peeling away the layers of secrecy surrounding the questions of the far-distant past. The complex of problems gradually unfolded before me, leading me bit by bit to more or less conclusive answers. We will again be leaving to posterity the hopes that they might engage with our research studies at the places where, with present day knowledge and techniques, they have been concluded. Thus the perpetuum mobile of the development of science continues revolving into infinity.

The rational scope of the book necessitated certain limitations, therefore the photographs of the skulls are not published in all projections, even though the frontal and basal norms would indeed be informative. Photographs of the skulls in all projections are obtainable from the author. Also, comparison of the cranial outlines is shown only in cases where the skulls are not heavily damaged. The research was limited by the lack of samples (the skulls available were only those which have been preserved); hence, for example, paleomicrobiological data on TB were not provided. Likewise, radiocarbon dating was not done, because its accuracy is not sufficiently high (+/– 30 years).

I wish to thank all of those who have trusted in me, have given encouragement and good counsel. Among so many experts, it is difficult to find for each the proper place: there are many whom I would gladly have named in first place!

I should like to express my warmest thanks to my colleagues from Ljubljana University, Slovenia, for offering me their advice: Professor of Forensic Medicine, Dr. Anton Dolenc and Dr. Branko Ermenc, Professor of Radiology Dr. Vladimir Jevtič, Professor of Anthropology Dr. Tatjana Tomazo Ravnik, and specialist in genetics Irena Zupanič Pajnič, MA, who were most generous with their advice.

The epigenetic research was conducted at the University of Vienna Medical Faculty. In these studies, I was unstintingly assisted by well-known epigenetic expert Professor Gertrude Hauser from the Histological Embryological Institute, and by Professor Johann Szilvássy from the Institute of forensic medicine.

For the genetic research, I was provided with information through the reference institutions. The work was conducted in the Paleogenetic laboratory of the Second Medical Faculty in Rome. I am most grateful to Professor Gian Franco De Stefano and Professor Olga Rickards.

The paleopathological research was carried out under the professional guidance of one of the most eminent European experts in paleopathology, Professor Eugen Strouhal at Charles University Prague, First Medical Faculty, Institute for the History of Medicine.

The X-ray examinations were performed in the Celje General Hospital. I should like to express my appreciation to the radiologist Dr. Mladen Malinar for having assisted me with his rich knowledge.

A special debt of gratitude is owed to the instigator of the research, Dr. Franc Štolfa; as a Celeian, he felt that the skulls of the Counts of Celje deserved to be scientifically studied. I am also thankful to medical historian Professor Edvard Glaser who supported my work all the time.

I am, furthermore, grateful to the staff of the Pokrajinski muzej (Regional Museum) Celje, Slovenia, where the skulls are preserved. The photographs of the skulls are the work of the local master photographer Viktor Berk. The study could also not have been accomplished without the consent of the Abbey of Celje.

In preparing the historical part of the research, I was fortunate to have good counsellors: the academician Jože Mlinarič, Professor Božo Otorepec, Professor Peter Štih and Rolanda Fugger Germadnik, MA.

For co-operation and personal involvement in the deliberations about the visual depictions of the Counts of Celje, I am grateful to the Academy painter, Rudi Španzel, for his remarkable portraits, drawn after the existing anthropometrical and anthroposcopical characteristics of the skulls.

For having rendered the entire discussion clearly and in fine English, I should like to express my gratitude to Alan McConnell-Duff, MA, and also for invaluable assistance in preparing and designing the text, to Marjetka Kočevar, MA, and to Metka Žerovnik, MA.

Finally, even though in last place – yet in emotional importance in first place – I wish to thank with especial warmth all those in my family for their understanding and full support.

CONTENTS

1.
INTRODUCTION

In Central-European and Balkan politics, the Counts of Celje played a more prominent role for just over half a century: the battle at Nikopol on 25 September 1396 marked the beginning of their meteoric rise to the stars – which, indeed, from the beginning of the 14th c. they had also borne on their coat-of-arms – while the Belgrade incident on 9 November 1456 tragically set the seal upon their fate. From then on, the exploits of this adventurous lineage, which in autumn 1436 was ennobled to princely status, continued to belong to history and to popular tradition. In subsequent periods, there was simply no limit that could be set to the imaginative reconstruction of the times that once had been.

Still today, the interest in the Celeians has not abated. Quite the contrary: the ever deeper subsiding of the Middle Ages into transience and the fabric of legend has granted to numerous interpreters of the past a constantly growing profusion of rather alluring material. The Counts of Celje have stirred the creative imagination of artists – painters, dramatists, storytellers and opera composers – and have also aroused the curiosity of scholars. Recollection of the Counts also marked the politically engaged writers, who wished to appropriate the Celeian memory to their own tendentious interpretations. During the 19th c., proponents of the German nationalist positions set out with exceptional ardour to prove the German-ness of the mighty feudal lords from the Valley of the Savinja; later, when the break-up of the Austro-Hungarian Empire led to the establishment of a completely new parallelogram of forces, they turned to expounding upon the Celeians' incorrigible Balkanism. In direct contrast to the latter, many Slovenes saw in the exploits of the Celeian dynasty the earliest endeavours to establish a common state of Southern Slavs. Meanwhile, the Romantic interpretations of the tales of »distant forebears«, during the period of integral nationalism, represented an important source of inspiration in shaping the visions of the future. Their anachronicity perturbed scarcely anyone. Yet the representations of a former »golden age« did enable many to survive the difficult present.

The Celeians, therefore, were – and remained – sufficient unto themselves alone. They did not allow themselves to be caught »without remains« in any explanatory mould. After them, it was possible only to inherit, but the true inheritor they never found. The political centre which they had shaped had completely died away. At the funeral of Prince Ulrich II in 1456, it was not in vain that the pathetic cry rang out: »Today the Counts of Celje, never more to be!« The anonymous chronicler recorded that these words were followed by indescribable outpourings of grief. Many now had come to realise that, hereafter, it never would be as it had been hitherto.

The almost pathological yearning for power, property and authority, which in the 14th and 15th c. characterized the frequently perilous pursuits of the Celeians on the Bosnian-Polish-Magyar-Czech chessboard, did not actually arouse any especial attention: in the »autumn of the Middle Ages«, throughout Europe, there were many noble kinsmen who wished to seize crowns, sceptres and thrones. Nevertheless, the fierce impetuosity with which the Celeians attained their aims diverged strikingly from the customary human measure. They were, clearly, well aware that political aspiration neither knows nor recognizes friends, but only sheer interests. Unscrupulously they schemed against the established order, and from the modest status of free Barons de Sovne (under this aristocratic title they first appeared in the records around the year 1130) they elevated themselves to become the powerful Counts of Celje. The title which »intabulated« them in history was conferred upon them on 16 April 1341 by the Emperor Ludwig IV the Bavarian, and in the next generation – in 1372 – this distinction was »renewed« in even greater splendour by his successor Charles IV of Luxembourg.

Family relationships with the Bosnian Ban-royal Kotromanić dynasty enabled the Celeians, in the second half of the 14th c., to present their sons and daughters on the European »marriage market«. Already in the first more decisive steps in this direction, Fortune smiled upon them. Count Wilhelm of Celje, on account of the far-reaching political calculations of his »relative« Louis the Great[1], became the husband of Anna of Poland. From their union a daughter was born, who carried the same name as her mother, in 1401 became the wife of Wladyslaw II Jagiello.[2] During the time of the fatal Battle of Grunwald she was occupying the Polish throne. She proved unable to charm her subjects, either with her beauty or with her eloquence, for – as can be gathered from the dusty medieval chronicles – she was as ugly as sin, and could only brokenly utter the language of her new homeland. Despite this, she did make a considerable impression upon her contemporaries – with her sound judgement and constancy of character.

The Celeians were everywhere afraid, and were scarcely anywhere accorded respect or even good favour. This, in the end, also led to their downfall: the magnates in the lands of the Magyar crown held them to be grasping foreigners, and in the Holy Roman Empire, during the wave of religious fanaticism – which north of the Alps burst out almost at the same time as the most splendid flourishing of the Renaissance in Italy – they were pronounced to be suspect religious fainthearts. The father-in-law of Emperor Sigismund, the powerful Count Hermann II, did at least outwardly still create the impression that he unreservedly supported the Holy Catholic Church and the papacy. His illegitimate son, whom he named after himself and ultimately also publicly acknowledged, became at first the Bishop of Freising and then the Tridentine Bishop. Hermann II was a skilful social climber, who, at the beginning of the 15th c. became Ban of Slavonia, Croatia and Dalmatia, and then endeavoured to »legitimize« his devotion to Christ's servants not only by establishing the monastery of Pleterje, where he was later buried, but also by his repellent religious-economic anti-semitism.

On the other hand, however, during the very same time the Counts of Celje had already agreed with the Serbian despot Đurađ Branković Smederevac that his daughter Catherine, after marrying Count Ulrich II (1433), would retain her Orthodox faith. Clearly, the spasmodic

1 The Magyar-Polish King Louis the Great married Elisabeth, daughter of the Bosnian Ban Stjepan II Kotromanić. Her sister, Catherine, was the wife of Wilhelm's uncle, Count Hermann I of Celje. King Louis, who after the death of Kazimierz III the Great acquired the Polish crown, married off the daughter of the former ruler to a rather unillustrious relative of his wife; thus he elegantly succeeded in removing forever any possible questions concerning the order of inheritance.

2 Wladyslaw II Jagiello, following the death of his first wife, Jadwiga (daughter of Louis the Great), endeavoured to consolidate his position on the Polish throne by marrying Anna of Celje, who was the granddaughter of one of the former rulers (i.e. Kazimierz III).

attempts to forge a church union between the Christian West and East reached a peak during the decades before the fall of Constantinople to the hands of Turks – had made no considerable impression on Herman and his counsellors. Likewise, the Celeians were not particularly interested in the crown in »heretical« Bosnia which, in 1427, had been promised to them by King Tvrtko II (if he were to die without heirs of his own).

Nevertheless, although a great many of Hermann's actions were subject to diverse judgements and commentaries – in particular, the contestable charter of his biography depicts the ruthless »elimination« of his unwanted daughter-in-law, the alluringly beautiful Veronika who, around the year 1422, »captivated« the firstborn Friedrich – he still remained a monumental figure in the consciousness of the populace of the late Middle Ages. His hand and his shadow were unnaturally long. This is why the »wrath« of numerous writers[3] was vented on Hermann's heirs. To many, they appeared to be dangerous religious nihilists. Prince Friedrich II was regarded as a sworn epicurean; some years before his death, when asked what use to his soul had been pilgrimage to the »Holy Place«, he allegedly responded – at least, so one may read in the renowned world chronicle of Hartmann Schedel (1493) – with the anthologically sarcastic catchphrase: »My cobbler goes on making boots, even after returning from Rome«. Even more degenerate, it seems, was his life-lusting sister Barbara, whose vital excessiveness brought many a grey hair to the head of her elderly husband, Emperor Sigismund of Luxembourg. There were several who saw in her a second Messalina, though one of the »more courteous« artists portrayed her as Venus on a galloping horse... One must, however, bear in mind that powerful personalities frequently arouse not only attachment but also opposition – and envy. This is particularly true of the exceptional women in any of the pronouncedly »masculine« eras. The Empress Barbara actually surrounded herself with military men and scholars who were able to appease her desire for learning and an unbridled testing of life's possibilities. She devoted particular attention to alchemy, while attributing no especial significance to the distinctions between the Hussites, the Utraquists and the Catholics. Consequently, many were persuaded that she did not care for the Germans (with the exception of Knight Walmerod).

Likewise, Prince Ulrich II was not regarded as a stereotypical 15th century man. For him – as also for his aunt Barbara – religious dogmas did not count for much: in his far-reaching political equations, even the Ottomans acquired their place. Ultimately, it was Sultan Murat II, brother-in-law of Ulrich's wife Catherine, for the Celeians and the Turks had a common adversary in the Magyar lands – the Hunyadi supporters. Yet, since long ago, enemies have made friends... Nonetheless, Prince Ulrich, without the aid of Allah's warriors, reached mainly into the lands of the St. Stefan crown (and even earlier into the Czech lands and Austria): he wagered on the card of Barbara's grandson László the Posthumous[4] – and in an instant even won the grand game. He became the prime political factor in the central Danubian region. In 1440 he played the main role at the spectacular coronation of his »nephew« László

3 Certainly the most important of them was Enea Silvio Piccolomini, who regarded the Celeians as devoted intellectual accomplices to Emperor Friedrich III.

4 The Emperor Sigismund of Luxembourg and Barbara of Celje had only the daughter Elisabeth, who married Duke Albrecht V of Habsburg. He later became Magyar and Czech King, although he died before the birth of his son László. Albrecht's second daughter Elisabeth was married to the Polish King Kazimierz IV.

in Székesfehérvár;[5] in 1452 he imposed himself as the boy's guardian, and in 1456 attained the position of regent in the Magyar lands. He tried to ensure peace for the country by adopting his most serious opponent, László Hunyadi – but the latter was well aware that there was room only for one at the top. He hatched a conspiracy against the Celeian and lured him into a trap in the fortress of Belgrade. Prince Ulrich II, knowing that he was the last of his lineage – his two sons and his daughter had already died, and his wife had become wholly alienated – after the perfidious blow from the Hunyadi supporters did die a knightly death.[6]

In the relatively extensive scholarly literature on the Counts of Celje, the study by Zvonka Zupanič Slavec presents an important new contribution: after the long decades of debate about the political, economic, cultural and religious implications of their appearance on the picturesque stages of late medieval history, this book takes us back, in the best possible literal sense, in medias res – to the very members of the dynasty themselves. Their existence is now reconstructed in the purely human dimension, which historiography – on account of its being directed towards the great supra-personal themes – frequently evades.

The times when the belittling judgement of the Celeians as being unscrupulous »Balkan potentates« seemed to be the final word on them are now forever relegated to the past perfect. Such most suggestively formulated stereotypical representations, nurtured by adherents of one cultural-civilizational circle about people from another, are sheer prejudices. Certainly, it is not possible to base on them any moderately adequate interpretations of times past. We know that the Celeians, for at least several successive generations, married almost equally as virtuously as did the Habsburgs. Through unusually ambitious politics they fixed their mark on their time – and it really was theirs. They performed the role of a kind of catalyser. In brief, they tolerated no indifference to their exploits. They stirred up passionate enmity, which reached its zenith with the expiry (male line) of the dynasty in 1456.

As patrons of the most diverse »chosen-ones of the Muses«, the Celeians were at least a factor of regional importance (in the area formed by the juncture of the Eastern Alps, the Pannonian plain and the Adriatic littoral). The humanist Johannes Rot(t), who later became the Lavantine and Breslau Bishop, even composed for Ulrich II a scholarly address in which he lauded the qualities of the assassinated prince.

So, too, it no longer remains unknown to us today that the Celeians were afflicted by the same diseases as were other nobles – and ordinary people. Once we have, one fine day, established to our surprise that on the peaks and in the valleys of the Balkans and in the neighbouring regions there was actually nothing different occurring than elsewhere in Europe (some centuries earlier or later), then we may begin to hope for at least an approximately reliable understanding of what the Celeians represented, and of their ideas, their exploits and defeats. The actual reality of the Middle Ages is certainly far more fascinating than either the black or the golden legends about them.

Prof. Igor Grdina, PhD

---

5 László the Posthumous was crowned while still an infant. A detailed account of this has been left by Helene Kottanner, the court lady of his mother Elisabeth. During the mass, which followed this ceremonial occasion, Prince Ulrich II held the holy Magyar crown above the child's head.

6 The elimination of the Celeians was of most advantage to Mátyás Korvín who, after the death of László the Posthumous (1457), ascended to the Magyar throne. He was Ulrich's unintended son-in-law (in 1451 Mátyás was betrothed to Ulrich's daughter Elisabeth) – and of course the brother of Ulrich's murderer.

2.
SCIENTIFIC HYPOTHESES

In this research, the following aims were established in advance:

1. to study from the **historiographical** aspect all available sources and all literature on the Counts of Celje. In doing so, it was hypothetically anticipated that in the registers and lesser known sources about the Celeians there would also be found several unknown or lesser known biographical details – about their diseases and the possible cause(s) of death, about their outward appearance, particularly the faces, and also about the burial places of individual members of the family;
2. to examine all the existing **genealogical** literature and to learn about whom one might expect to find among those communally interred;
3. to verify the **hypothesis that the skulls did actually belong to family members of the Counts of Celje**. Following the prior studies, it was expected that a specific number of male, female and child skulls would be found;
4. to study from the **somatic aspect**, the 18 preserved skulls, to determine gender and age, diseases and injuries, degenerative changes and possibly even the causes of death;
5. to establish the **identity** of the skulls studied and the kinship relations between them. In the hypothesis, it was foreseen that the skulls would be hereditarily linked from Friedrich I to Ulrich II of Celje;
6. to **clarify the principle of numbering** of the extant skulls. Hypothetically it was anticipated that the numbering would not follow the burial chronolgy of the family members;
7. to study the injuries to **skull 18** by using forensic medical expertise. Both hypothetically and in accordance with historical sources, the skull has been attributed to the Count, later Prince Ulrich II;
8. to study the pathology on **skull 14**;
9. to establish the **shaping features** of the heads and faces;
10. to introduce the **identificational method for determining family linkage** in cases where only the skull from the communal vault is available, and it is anticipated that there will be a blood relationship;
11. to compare the historiographic, genealogical and somatic findings and to incorporate them into the new revelations of this interdisciplinary research.

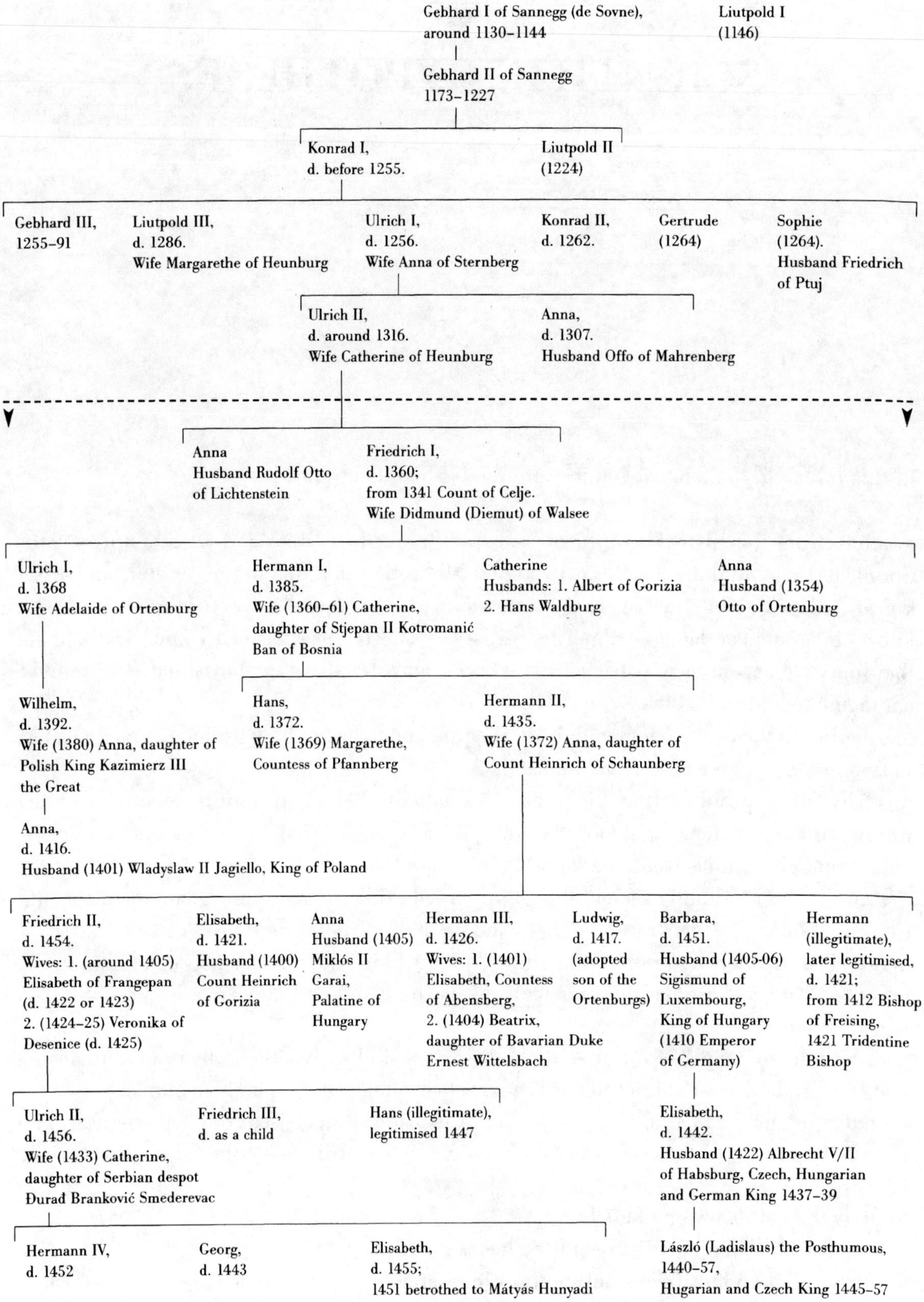

Picture 1: *Genealogical tree of the Counts of Celje (Enciklopedija Slovenije, 1988). The counts from Friedrich I onward buried in the family vault in the time period between 1354 and 1456 (indicated by arrows)*

3.
METHODOLOGY OF WORK

3.1.
Historiographic Methods

Historiography provides the framework for this study:

it presents the personal features of the family of the Counts of Celje, and records the available biographical and genealogical information. In addition, it deals with the question of interment in the family vault at the Celeian Minorite church. An examination was made of all the registers, transcriptions and extracts from the documents of the Celeian Counts in Latin and Old German, all the secondary sources, all domestic and foreign books, and important articles which had been published about them.

Probably the contention of the Slovenian paleographer, Prof. Dr. Božo Otorepec, will hold true, i.e. that, after the demise of the Celeian family line, the Habsburgs examined their archives and preserved only those documents that were of importance for political, economic and other reasons, but not those of family concern. Among the extant documents relating to the Celeians – which Slovenia, after independence, succeeded in acquiring from various Austrian archives (about 850 documents) – there are almost none which relate to inheritance, nor are there any testaments, documents on the birth or death of individual family members, nor any records of burials. Only a few, rare records have been preserved which make indirect reference to the death of individual family members, widowhood, betrothals and marriages. Since it was not possible properly to reconstruct the birth and death data from the original documents and registers, we drew upon the extant records from the Chronicle of Celje and upon other secondary sources, which occasionally also present contradictory assertions.

3.1.1.
History of the Counts of Celje

Concerning the Counts of Celje, much variegated literature has been written, containing information which is frequently not compatible, at times even contradictory. For the present study, the economic and political considerations were of lesser importance; nevertheless, it was still necessary to consult all the extant sources and literature in order to approach the existing biographical, genealogical and chronological information on the Celeians, and also to gain access to the rare birth-records of individuals and to the somewhat more frequently mentioned death-dates. Furthermore, of course, also to be able to test out the conjectures and conclusions concerning the persons who might actually have been interred in the Celeian Minorite church.

The main source concerning this family is a Chronicle by an unknown author – probably a Celeian Minorite monk – who was well acquainted with the family of the Counts, and recorded it for the period extending from 1341 to 1456. This Chronicle is preserved in 17 transcriptions, written in Middle German. The Chronicle was translated into Modern German, with a historical and critical analysis, by Franz Krones, and published in Graz in 1883 (Krones 1883). It was later translated into Slovene by M. L. Golia, 1972. In the mid-19th c., I. Orožen again drew attention to the noble family in the Chronicle of Celje (Orožen 1854, 1997). Among the older authors, important information was also contributed by A. Gubo (Gubo 1909) and G. May (May 1943), and by H. Dopsch (Dopsch 1974/75). Attention to Venetian sources has been drawn by the scholar from Trieste, F. Cusin, with his publication of material on Celeian politicians in the North-Adriatic hinterland (Cusin 1937, 1977). Within the context of Slovene history, others who wrote about them include J. Gruden (Gruden 1910, 1992), M. Kos (Kos 1933) and, most extensively, J. Orožen (Orožen 1927). Krones' analysis of the contemporary sources on the Counts of Celje has revealed that the most important information comes from the pen of E. S. Piccolomini, the **Pope's envoy to the court of Emperor Friedrich III**, an ingenious humanist, later to become Pope Pius II; he described the importance of the Celeians with regard to central European political events, and also characterized them personally (Baš 1951, 1989). Since he was politically of pro-Habsburg inclination, and hence not favourably inclined towards the Celeians, in his writings he tended more to stress their negative sides, as did the Magyar and Czech sources, which mainly treated them as foreigners. They are reflected in a different light, however, by the Chronicle of Celje, the historiographically most important source about them. It was written by their loyal adherent, mainly in memory of counts Hermann II, Friedrich II and Ulrich II. The Chronicle was first edited for the Slovenes by Ignac Orožen in 1854.

The many contradictory sources on the Counts of Celje, and particularly the uncertainty about the death of Elisabeth of Frankopan and of Veronika of Desenice, and also about the sudden death of Ulrich II, have led to highly differing literary and musical depictions of this theme. In order to confirm the information about the Counts of Celje it is useful to compare the earlier mentioned, most exhaustive, yet otherwise conflicting sources – Piccolomini's work and the Chronicle of Celje.

The ascent of the Counts of Celje began in 1341, when the nobles from the Sannegg family became Counts of Celje, and ended in 1456, when Prince Ulrich II was assassinated by conspirators under the directives of Ladislaus (László) Hunyadi in Belgrade. The rule of the counts was contemporaneous with the emergence and fall of the Hussites, the birth of Switzerland, the settlement of the Turks on the Balkan peninsula, and the conquest of Constantinople in 1453. Their times also saw the discoveries of printing (around 1440) and of gunpowder; the great Gothic cathedrals of Prague and Seville were being constructed, and the universities of Prague and Vienna were established. The Counts of Celje were contemporaries of the Italian Renaissance, the religious reformer Jan Hus, the French heroine Jeanne d'Arc, and the Florentine ruler Cosimo de Medici. At the time of the demise of the Counts of Celje the wars of the Crusaders were ending in Europe, and Leonardo da Vinci was born.

On the Balkan Peninsula, Byzantine feudalism was crumbling under the pressure of Turkish military might. New states were beginning to emerge which were no longer founded exclusively on feudal legitimation and inheritance, but were being shaped by strongly talented individuals, properly ordered military forces, and economic interests. This led to a loosening of the medieval unity of the Holy Roman Empire and the Church of Rome, and to a weakening of the influence of castles and monasteries.

Within the framework of the history just described, F. Baš (Baš 1989) divided the political development of the Celeians into three periods. During the first, the Sannegg family in 1322 inherited half of Celje, because they were related by kinship to the Heunburg family, which had died out, and in 1341 were elevated to the status of Counts of Celje. They then took over the coat-of-arms of the noblemen of Heunburg, with its three gold stars on a deep blue field.

Coat-of-arms of Hermann and Friedrich of Celje (Das Concilium zu Constenz 1536, page CLVIIIb)

During the second period, the Celeians developed politically by bypassing the Habsburgs and linking their interests with the nobility of Luxembourg. Through them, they gained a foothold, both politically and in terms of property, in Hungary and Croatia. With their landholdings over the Croatian Zagorje and Medjimurje, and their custody over the Zagreb diocese, they became one of the first great estate owners in Croatia, and through inheriting the Ortenburg properties in Koroška (Carinthia), Gorenjska (Upper Carniola) and Dolenjska (Lower Carniola), they also became the first political force in the Slovene lands. This was also legislatively confirmed by the Emperor of Luxembourg in 1436, when he elevated the Counts to the title of Prince. The cause for the new direction taken in Celeian policy was the weakening of the Habsburgs – on account of the subdivision of their lands into the Austrian and Inner-Austrian group in 1379 and, in 1411, the new division of the latter into Tyrolean and Inner Austria, with Trieste, Istria, Kranjska (Carniola), Koroška (Carinthia) and Štajerska (Styria). The marriage of Friedrich II to Elisabeth of Frangepan added to the Celeians' possessions Bakar with its environs, and hence access to the Adriatic. For dominance over the sea, the Venetians were also fighting against the Magyars. Their anti-Habsburg and anti-Venetian interests drew the Celeian counts into the imperial Luxem-

Celje, Schedel, Buch der Chroniken 1493, page CLXXXIII

bourg camp, which actually had its political gravitation point in Hungary (Baš 1989).

During the third and final period, after the year 1436, the Celeians were fighting for supreme authority over Hungary, and through inheritance contracts with the Habsburgs in 1443 and the Walseeians in 1451 they extended their territory to the south through the regions of Dolenjska (Lower Carniola), Bela krajina (White Carniola), Kras (the Karst), Istria and Rijeka (Fiume) to the Adriatic.

Hermann II was the most consolidated political personality among the Counts of Celje. In both domestic and foreign policy affairs he linked the Celeian interests with the reality of the Church of Rome and the Germanic Empire. As is revealed by the monasteries at Pleterje in Dolenjska (Lower Carniola) and at Lepoglava in Zagorje, he was also closely linked to the artistic life of the time. At the court of Celje, Hermann II of Celje was the absolute ruler. In terms of the economy, he conceived the Celje estates as being transit territory for transport from the Venetian province to Austria and on to Hungary, and also directed the import of sea-salt to Celje; then, through the marriage of his son Friedrich II of Celje to Elisabeth of Frangepan, he opened the way for the Celeians to have access to the Adriatic. In the 14th and 15th centuries, the natural economy was being increasingly replaced by the monetary, and consequently the Celeians acquired supervision over mining activities and monetary matters, thus consolidating their financial strength. By expanding his economic and political power, Hermann II also surpassed religious ways and prepared the circumstances for arranging the marriage between the (Slav) orthodox Catherine Branković and his nephew Ulrich II. Branković, as a Celeian countess, remained orthodox and had her own two orthodox priests at the court. Their two sons became catholics, while the daughter remained orthodox.

In complete contrast to Hermann's powerful ruling personality was that of his son, Friedrich II (born in the 1380s, died in 1454, and ruled from 1435 to 1454); he was the true »seigneur of the Celeian dynasty, and lived a life of full self-indulgence« (Baš, 1989). The difference between father and son with regard to politics was soon observed by the Venetians, who reappropriated the patriarchate of Aquileia, and instead of relying on the more than 60-year-old Friedrich, counted rather on his 30-year-old son Ulrich II, who had completely overshadowed his father politically. Historians also frequently compare Friedrich II with similar personalities of the time, e.g. Cesare Borgia, or else characterise him as a half-Hamlet, lacking the true willpower to realise his rulership aims by means of his own actions (Habjan 1988). His love life, together with the family drama of his first wife Elisabeth of Frangepan and his mistress Veronika of Desenice, caused the Celeian despot to squander the access to the Adriatic, and prompted his opponents among the Frangepans to form an alliance with the Venetians against him. Friedrich was lacking in both political and dynastic sensibility.

The last representative of the Celeian counts, elevated to princes, was the son of Friedrich II, Ulrich II (1406–1456). He was the most ambitious politician within the Celeian dynasty. Piccolomini, in a letter of 1445 to Bishop Lenart of Passau, described him as a sensible and intelligent man (Baš 1989) with a decisive bearing. The Chronicle of Celje portrayed him as a knight who worked for the good of his King Ladislaus the Posthumous and for the Catholic church. Following his uncle's death, he effectively led all Celeian politics. He was, in fact, not so much the representative of the dynastic territories but was, above all, the deputy of Albrecht of Austria, the husband of his niece

Elisabeth, as his regent in the Czech lands (1438–1439), and then the deputy of his son Ladislaus the Posthumous in Austria (1452–1453, and again in 1454). Following the death of János Hunyadi in 1456, he became the royal regent in Hungary. In order to realise his aspirations, he had to fight in the Czech lands and in Hungary against those classes who were opposed to foreigners. As a foreigner himself, he was fighting for authority in those parts, and in doing so it would appear that he lacked political realism. This was, at the same time, the last act of the Celeian political drama and of its feudal coloniser, Ulrich of Celje. All of his three children had predeceased him. His violent death at the hands of the conspirators of Hunyadi from the Hungarian court marked the almost anticipated end of the last of the Counts of Celje, Ulrich II.

Barbara of Celje, Emperor Sigismund's wife (Schedel, Buch der Chroniken 1493, page CCXLII)

The historical survey of the pillars of the Celeian family should not represent only the male members. Among the female members stands out the powerful figure of Barbara of Celje, daughter of Hermann II and wife of King Sigismund I of Luxembourg. In the literature, she is usually compared to Lucreza Borgia. If we wish to understand Barbara's role, we must be aware that in the medieval courts the wife's task was to take care of the children and supervise the domestic staff, while the knight husband spent the greater part of his time on campaigns and in hunting. The same medieval marriage conditions also prevailed between Friedrich II and Elisabeth of Frangepan, and between Ulrich II and Catherine Branković. The Renaissance reintroduced orderly family life, and this was a new-age phenomenon. Barbara of Celje is held to be a woman with a wide-ranging, free-thinking mind. She was involved with the modern sciences of the time, astrology and alchemy. Barbara was religiously tolerant, and co-operated with the Czech Hussites; it was they who buried her after her death. She also engaged in the political arena, and it was this involvement that led to her temporary banishment.

Friedrich II and Ulrich II lived at a time when Italy was overwhelmed by humanism and the Renaissance, however they – like many other noble persons – did not live to experience it. Barbara had been caught up by the new spirit abroad, and she accepted it.

3.1.2.
From Biographies of the Counts of Celje

The initiator of the Celeian lineage was Gebhard de Sovne (1130–1144), who lived in the castle of Sannegg (Žovnek) in the Valley of the Savinja river. They intensively extended their estates, so that after the year 1333 Friedrich of Sannegg was the sole owner of Celje and its overlord; then, in 1341, **the Sannegg family converted their name to** of Cilli (nowadays of Celje).

Friedrich I of Celje acquired the estates by purchasing or by lending money on pledge. Like the Jews, he loaned money on a large scale, often acting as creditor or guarantor; his guaranty often brought him a castle or other property.

The extensive estates of Friedrich of Sannegg aroused in him the desire to acquire the title of Count. This was accorded to him by Ludwig of Bavaria on 16 April 1341, in München. Thus the men of rank of Celje became the Counts of Celje.

Friedrich I was married (see genealogical tree) and had two sons, Ulrich I and Hermann I, and daughters Catherine and Anna. Their birth details are not known. Ulrich I was presumably born before 1330. In 1345, it is recorded that he was engaged in battle around the Dalmatian city of Zadar, and one year later in the Tyrol. In 1350, he was in Prussia and Lithuania; in 1354, he accompanied the Emperor Charles IV and his army to Rome, again a year later he was fighting with the Croatian-Hungarian army against the Venetians, and in 1365 in Bulgaria. He was married to Adelaide of Ortenburg (no biographical details available), with whom he had a son Wilhelm (d. 1392). Wilhelm in 1380 married Anna, daughter of the Polish King Kazimierz III the Great; they had a daughter, also named Anna (d. 1416), who married the Polish King Wladyslaw II Jagiello (1401). Ulrich I died in 1368.

The second son of Friedrich I, Hermann I (d. 1385), joined his brother in various affairs of ruling only after their father's death in 1360. By 1361, Hermann I was already married to Catherine Kotromanić, daughter of the Bosnian Ban Stjepan II Kotromanić, a descendant on the maternal line of the Serbian Nemanjić family. They had two sons, Hans (d. 1372) and Hermann II (d. 1435). Hans was betrothed as a child to Countess Margarethe of Pfannberg (1369), but died in adolescence. His brother, Hermann II, was married (1372) to Anna, daughter of Heinrich of Schaunberg; they had six children in their marriage, but Hermann II also had two illegitimate children – a son (Hermann) and a daughter.

In 1377, an **inheritance contract** was drawn up between the Celeians and the Ortenburgs, just prior to the departure for the war of the Crusades: if the Celeians were to die out without male

The Castle of Sannegg, Vischer, Topographia Ducata Stiriae 1683

descendants, then the main inheritors would be the Ortenburgs, and vice versa (Orožen 1927).

The Countship that had been conferred upon the Celeians by Ludwig the Bavarian in 1341, was valid only for certain feudal estates. Ludwig's successor, Karel IV of Luxembourg, in 1348 pronounced all of Ludwig's decrees, which might be injurious to the Habsburg estates, invalid. However, since Ulrich I and Hermann I had both maintained good relations with Karel IV, on 30 September 1371 he reinstated their title as Counts, i.e. for the County of Celje, comprising the upper and central regions of the Posavje (Štih 1996). Hermann I died on 31 March 1385. He was laid to rest in the family vault of the Celje Minorite Church.

Hermann **II of Celje** initially ruled together with his cousin Wilhelm (d. 1392), but after 1390 became regional princely ruler of Kranjska (Carniola).

At the end of the 14th c., Central Europe was frequently menaced by the danger of Turkish incursions. In the year 1396, Hermann II engaged in the war against the Turks, and in battle saved the head of King Sigismund of Luxembourg. On another occasion, he also rescued him from imprisonment. As a mark of gratitude, King Sigismund betrothed himself to his daughter Barbara, although on account of her youth the marriage did not take place until 1405 or 1406. The alliance with Sigismund greatly elevated the reputation of Hermann II, and furthermore, his future son-in-law presented him with numerous estates. The illegitimate son of Hermann II, likewise named Hermann, later became Bishop of Freising, and was granted in pledge the Loka estate for a period of four years (1417–21) (Grafenauer 1988).

When King Sigismund married Barbara of Celje, as a generous son-in-law he presented his father-in-law with Medžimurje and Čakovec, and Hermann II became the Slavonian Ban, with the right to appoint the Bishop of Zagreb. Thus he became the most eminent noble of the Magyar-Croatian crown. The new estates reached from the Savinja river to the Sotla river. On 30 June 1436, one year after the death of Hermann II, Friedrich II and Ulrich II were elevated **to the title of Prince** in Prague, as it is recorded in the document (Štih 1996).

Hermann II died on 13 October 1435 (Grafenauer 1988) in Bratislava, shortly after the elevation of the Celeian counts to the title of Princes. When King Sigismund elevated Friedrich II and his son Ulrich II to the rank of state princes in

Prague he was at the same time founding the Celeian princedom; it comprised the counties of Celje, Ortenburg (Gruden 1910) and Strmec (Počnik 1909). With the acquisition of this status they ranked equal to the Habsburgs. For the Habsburgs their territories were becoming an increasingly serious obstacle to the route towards the sea; it is therefore not surprising that this led to severe tensions between the two dynasties, and to a long-lasting battle, which was concluded in 1443 with the signing of several contracts, one of which was an inheritance agreement. In a special privilege, Friedrich III of Habsburg elevated Friedrich II and Ulrich II to the rank of Prince (Štih 1996).

Hermann II incorporated the qualities of governing decisiveness, power and unscrupulosity, he had a notable political acumen and a developed vision of the future of the Counts of Celje, yet he was also a man of patience.

Friedrich II, son of Hermann II and Countess Anna of Schaunberg, grew up alongside his father's powerful personality. His nature was not up to the great political mission which had been intended for him. While he was still a child he was given by his father in betrothal to Elisabeth of Frangepan, the only daughter of Stefan I of Frangepan. They were married around the year 1405. In 1406 their first child was born, the son Ulrich II, then followed Friedrich III, who died as a child, and Friedrich's illegitimate son Hans, who was later acknowledged in 1447. Friedrich II and Elisabeth of Frangepan first lived together at the castle in Krško, but then he left her and carried on his passionate, unbridled life. He became acquainted with Veronika of Desenice when he was already aged about fifty. On account of his great desire to marry and live with her, and because of his father's resolute opposition and insistance that Friedrich return to his legal wife, according to the general hearsay he murdered Elisabeth of Frankopan (in 1422 or 1423). The father punished his son's wilfulness by having him imprisoned and also by depriving him of all his estates. The marriage to Veronika of Desenice did not end happily. Hermann II gave orders for her to be killed in 1425.

During the remaining almost thirty years of his life, Friedrich II did not contribute to the political and economic might of the family. His role was played, instead, by his son Ulrich II. Among all the Celeians, Friedrich II probably attained the greatest age, dying in his eighties, in 1454 (according to Krones, on 20 July 1453), i.e. just two years before the demise of the male descendants of the family line.

Since Friedrich II had no political acumen, Hermann II rested his hopes upon his grandson **Ulrich II**, who continued to pursue the political work of his grandfather. He was comparable to Hermann II in his indestructible will, energy and outward appearance. In 1433, his grandfather had arranged for him to marry Catherine Branković. At first, harmony reigned between the couple, but later Ulrich II became wearied of his wife and left her. She had born him three children: Hermann IV, d. 1452 according to Grafenauer, or 1444 according to Cusin; Jurij (Georg), d. 1443 according to Grafenauer, while according to Krones he had already died in the cradle on 1 May 1424 (Krones 1972); and Elisabeth who, according to the data of Dr. Christiane Thomas (Thomas 1991) and her research studies of 1991, was born between the years 1443/44, and died in the autumn of 1455. On 1 August 1455, she bore a son to Ladislaus (László) Hunyadi. Concerning Hermann IV, the same author in her genealogical tree gives the date of his birth as 1432/33, and death as 30 July 1444. All three of Ulrich's children would, therefore, have died in early adolescence.

Ulrich II continued to actively pursue his political path after his grandfather's death. He maintained residence mainly at the court of his uncle, Sigismund of Luxembourg, who was not only the Magyar, German and Czech King, but also Emperor of the Holy Roman Kingdom. He also elevated Friedrich II and Ulrich II to the status of prince. The decree on princely ennoblement was issued in Ulrich's presence in the Old Town of Prague on 30 November 1436 (Orožen 1927). In 1437, King Sigismund of Luxembourg died. His daughter Elisabeth (daughter of Barbara of Celje), Ulrich's cousin, was in 1422 married to Albrecht V/II of Habsburg, who, following Sigismund's death, became Czech, Magyar and German King (1437–39). The coronation of his son – still an infant – as Ladislaus the Posthumous was held after Albrecht's death, on 15 May 1440. He was Magyar and Czech King from 1445 to 1457; on 23 November 1457 he suddenly died (of poisoning?).

Celje, Vischer, Topographia Ducata Stiriae 1683

After the death of all three of his children, Ulrich II was left without descendants, and so he endeavoured to adopt Hunyadi's oldest son Ladislaus (László), although without success. Under the leadership of Ladislaus (László) Hunyadi, the Hungarian nobility had been plotting the assassination of Ulrich II, because at a state council assembly held in early November 1456 at Futog near Petrovaradin, Ladislaus the Posthumous had designated Ulrich II as supreme commander and state regent in Hungary.

Consequently, when Ulrich II was at the height of his powers, the Magyars hatched a conspiracy against him. During the joint campaign against the Turks, on 9 November 1456 they had him decapitated (Krones 1972). His corpse was transferred back to Celje, where it was interred in the family vault. Ulrich's widow, Catherine Branković, for many years firmly resisted against yielding their estates to the Habsburgs, as had been agreed in the inheritance deed. After being defeated, she set off to rejoin her sister Mara in Macedonia, and died around the year 1490. She is buried at Konća, west of Strumica (Krones 1972). It is interesting to note that her sister Mara was married to the Turkish Sultan Murat V (Krones 1972).

And thus the family line of the Count Princes of Celje unexpectedly expired with the death of Ulrich II, for he had remained without male heirs.

Pleterje, Valvasor, Die Ehre des Herzogtums Krain, XI, 444

3.2.
Identification Methods

3.2.1.
Use of Identification Methods for Determining Gender by Skulls

The outer surface of the skull is influenced by life age, body height, race, individual development, and environmental factors. Between the ages of 20 and 55 years, the appearance of the skull remains almost unchanged; earlier it is influenced by puberty, and later by senium. Prior to puberty, there is almost no difference in shape between male and female skulls.

The most important gender differences in skulls, as established by Hunger and Leopold (1978), are in their size, glabella, supraciliary arches, orbital openings, frontal tubers, frontal grooves, supraorbital margins, inclination of the forehead, cheekbones, occipital relief, external occipital protuberance, and mastoid processes. Rarely does one encounter skulls which might have all the theoretically determined features of inter-gender differentiation. Consequently, for reliable determination of gender, one should always survey all the theoretical suppositions.

Upon first viewing, the male skull is larger and more robust, while the female skull is generally narrower and more gracile. After the second decade of life, the glabella becomes pronounced in the male, while in the females this is almost never observed. In males, the mastoid is notably larger and more robust (Hunger, Leopold 1978). In males, the great occipital foramen is both absolutely and relatively larger than in females (Hunger, Leopold 1978). The facial bones in males are more strongly developed and also larger in absolute measures.

Keen (Keen 1950) recommends four measurements for gender determination: measurement of the greatest cranial length, the distance between the cheekbones – bizygomatic breadth, width of the mastoid and depth of the fossa infratemporalis. In addition, he also recommends taking into account three morphological characters: supraciliary arches, occipital relief, and the upper external edge of the acoustic meatus. With these seven features, Keen was able to determine gender in 85% of cases.

Stewart (Stewart 1954) and others are of the opinion that it is possible, simply by inspection of the skull, to reliably determine gender in 80–92% of cases. For sexual dimorphism in skulls, the following measurements are most important: cranial length, bizygomatic breadth, and proportions between the width and height of the orbits. Also important for determining gender are the mandibles, which in our case, however, are not preserved.

Ascádi and Nemeskéri (1970) recommend several parameters for description of the skull: glabella, mastoid process, plane of the nuchal area, supraciliary arches, frontal tuber, external occipital protuberance, and inclination of the forehead. Among the measurements: for the neurocranium determination they propose the largest cranial length and width, and the cranial bases; and for the splanchnocranium the orbital height, left and right, the width of the anterior nasal aperture and the bizygomatic breadth (picts. 2, 3, 4).

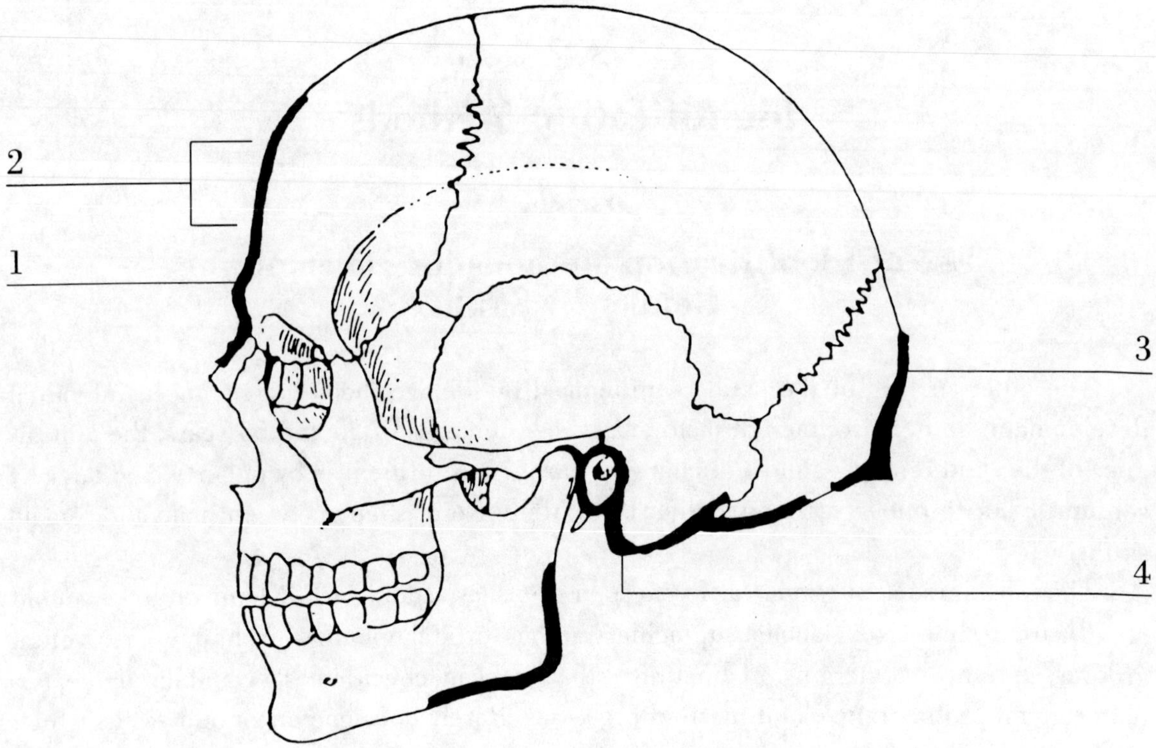

Picture 2: *Characters for determining gender by skull, side view (Ascádi, Nemeskéri 1970): 1 glabella, 2 inclination of the forehead, 3 external occipital protuberans, 4 mastoid*

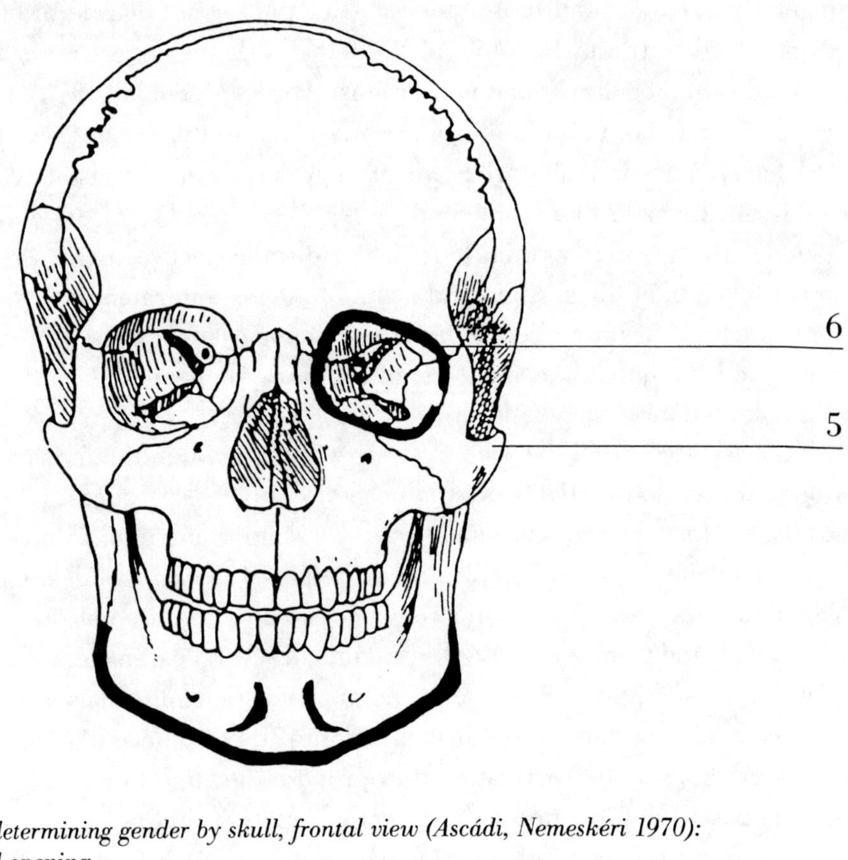

Picture 3: *Characters for determining gender by skull, frontal view (Ascádi, Nemeskéri 1970): 5 zygomatic bone, 6 orbital opening*

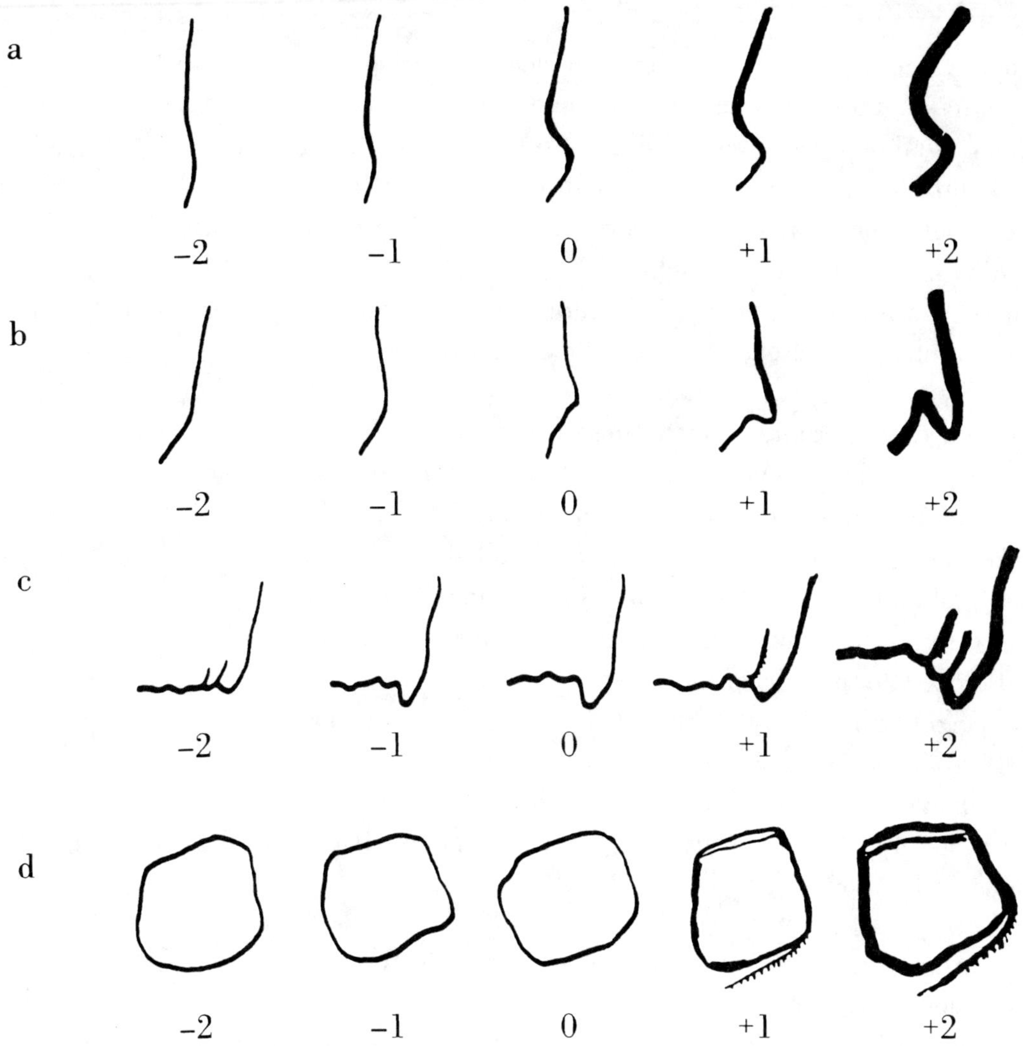

Picture 4: *Assessing gender after Ascádi and Nemeskéri from (−2) to (+2): a) glabella; b) external occipital process; c) mastoid process; d) orbital shape.*
Key to the numbers: (−2) – hyperfeminine; (−1) – feminine; (0) – neutral; (+1) – masculine; (+2) hypermasculine

Morphological characters, combined with craniometry, improve the possibilities of correctly identifying the skulls by gender. If, by using these identification methods, one is unable to reach final conclusions, then radiographs and forensic dentistry are used as an aid. The procedures for determining gender by skull are, therefore, based on morphological characters and on anthropometric and dental methods.

3.2.1.1. Morphological characters for determining gender

(Hunger, Leopold 1978; Milčinski 1958; Martin, Saller 1957; Dominok 1959; Schwidetzky, 1969; Keen 1950)

- superciliary arch: in males (m.) strongly developed, in women (f.) weakly developed;
- glabella: in m. after the age of 20 becomes even deeper and stronger, in f. it is unpronounced;
- the upper external margin of the acoustic meatus: more pronounced in m.;
- occipital relief: more pronounced in m. than in f.;
- maxillary palatine process: more ellipsoid in f., in m. more rounded;
- forehead profile: more inclined in m. than in f.;
- cranial relief from behind: in m. more pronounced, prominent places for muscle attachments, in f. the relief is slighter;
- thickness of calvaria: in m. occipitally larger than frontally, in f. over the age of 40 stronger frontally; (Rämsch, Herrmann 1963/64);
- frontal tubers: stronger in m. than in f.;
- muscle attachments: more powerful in m., slighter in f.;
- facial part of skull: in m. larger and broader, in f. smaller and narrower;
- dental arch: in m. more rounded, in f. more elliptic;
- orbital entrance: in m. angular, in f. more rounded;
- supraorbital margin: in m. thicker and more rounded, in f. thinner and sharper;
- size of teeth (particularly molars): in m. on average larger, in f. smaller;
- styloid process: in m. thicker, in f. thinner;
- orbital height: in m. slightly higher than in f.;
- orbital breadth: in m. considerably broader than in f.;
- nasolacrimal duct: deeper in m., shallower in f.;
- attachments of the temporal muscle (fossa infratemporalis): in m. deeper and more spacious, in f. flatter;

3.2.1.2. Determining gender by anthropometric methods and indices

(Martin, Saller 1957; Trimble 1987; Hrdlička 1939; Smodlaka 1948)

- **greatest cranial length (between the glabella and external occipital protuberance):** in m. 175–193 mm, in f. 160–182 mm;
- bizygomatic breadth (between both zygiones) m. 128–143 mm, in f. 120–135 mm;
- greater occipital foramen (the greatest length is measured along the sagittal axis, and the greatest breadth along the medial axis): in m. absolutely and relatively larger than in f. The antero-posterior measurements are longer in m., and the shape is more ellipsoid, in f. the foramen is more rounded; the length in m. is 36–41 mm, in f. 30–35 mm, the average breadth in m. is 30.6–35 mm, in f. 25–30.8 mm;
- mastoid process: its breadth is measured, in m. 28–32.5 mm, in f. 25–28.7 mm;
- cranial weight: in m. average 720 g, in f. average 590 g. This method can be used only on fully preserved skulls;
- cranial capacity: in m. average 1450 cm^3, in f. average 1300 cm^3;
- cranial base: in m. broader, 97–109 mm, in f. narrower, 90–95 mm.

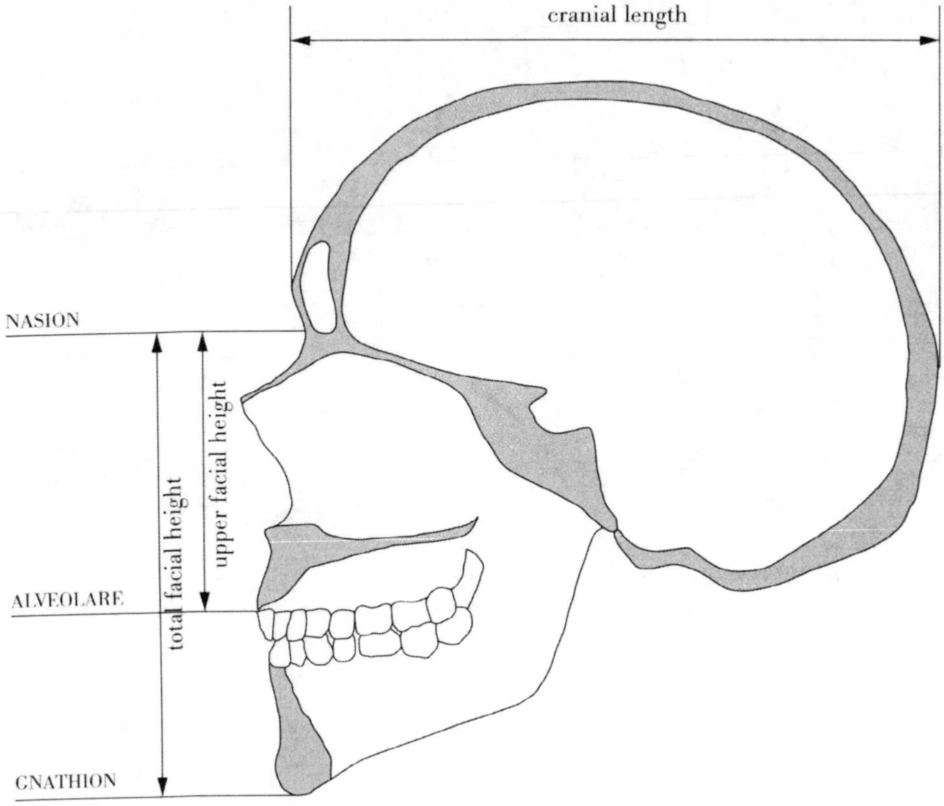

Picture 5: *Craniometric points for determining the total and upper facial indices (Brass, Trimble 1987). The total facial height is measured by the distance between the nasion and gnathion points, and the height of the upper part of the face by the distance between the alveolare and gnathion points*

Anthropometric indices on the skull

(Martin, Saller 1957; Trimble 1987)

The **cranial index** is used as an aid in determining the gender and shaping features of the head – the proportions between the breadth and length of the skull, multiplied by 100. In accordance with this index, skulls may be distinguished as being narrow or long, average length, and round or short (Trimble 1987):

- dolichocephalic – long (X–74.99)
- mesocephalic – average length (75–79.99)
- brachicephalic – short (80–84.99)

Concerning the shape of the face, much is conveyed by the **facial index** (pict. 5):

the proportion between the height and breadth of the face (bizygomatic breadth) multiplied by 100.

Using this index, one may determine the three basic forms of the face: broad, medium, and narrow (Bass, Trimble 1987):

- euriprosop – broad (80–84.99)
- mesoprosop – medium (85–89.99)
- eptoprosop – narrow (90–94.99)

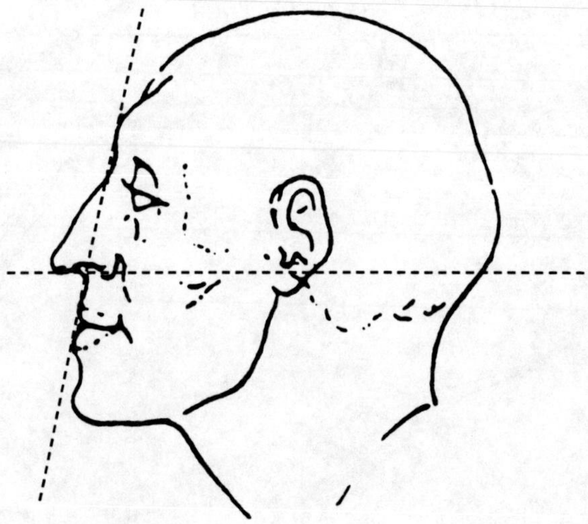

Picture 6: *Orthognathy after Krunkenberg – Campers' angle is almost straight*

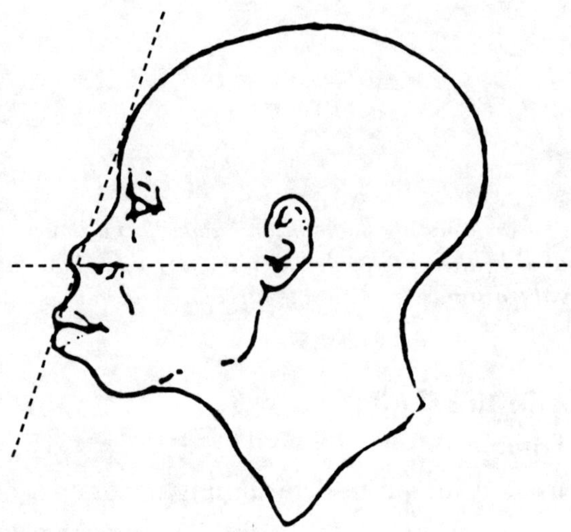

Picture 7: *Prognathy after Krunkenberg – Campers' angle is sharp*

Since the mandible was missing it was not possible to calculate the total facial indices in this study, however it was possible to calculate the **upper facial index**. This represents the proportions between the upper height of the face (the distance between the alveolar and nasion points) and the breadth of the face (bizygomatic breadth), multiplied by 100.

Retzius (Bikar 1970) was the first to undertake craniometric analyses. He measured the **angle** between the forehead and the labial surface of the front and upper incisors, and by means of this line the orthognathy or disgnathy. He used the so-called **Campers' angle**; this is enclosed by two lines, one of which runs from the frontal protrusion to the labial surface of the incisors, while the other links the aperture of the outer auditory canal with the front nasal spine. Using these lines he studied facial profiles.

In relation to the size of Campers' angle, Retzius described the orthognate and prognate skulls. Campers' angle is also important for studying the skulls which are supposedly those of the Counts of Celje: it determines the shape of the upper part of the skull, since fragmentarily preserved skulls without the mandible, and measurements without additional theoretical aids, do not permit conclusions to be drawn about the original appearance.

If the labial surface of the upper front teeth is approximately at a right angle to the frontal protuberance, we speak of orthognathy after Krunkenberg, while if the angle is sharp it is referred to as prognathy after Krunkenberg (picts. 6, 7).

In this study, the following indices have been observed:

- cranial index:
  I = (maximum cranial breadth / maximal cranial length) x 100, which gives information on the cranial shape (dolichocephalic, mesocephalic, brachycephalic);
- total facial index (pict. 5):
  I = (total facial height x 100) / bizygomatic breadth, which gives information about the shape of the face (euriprosop, mesoprosop, leptoprosop)
- upper facial index (see scheme):
  I = (upper facial height x 100) / bizygomatic breadth, which gives information about the upper part of the face;
- orbital index (Bass, Trimble 1987):
  the breadth and height of the orbit differ significantly by gender.
  I = (orbital height x 100) / orbital breadth
  in females the average index of the orbit is 90.07, in males 88.03;
  (Hunger, Leopold 1978)
- Campers' angle is bound by two lines: the almost horizontal line between the two apertures of the outer auditory canal, and the line between the frontal protuberance and the labial surface of the upper incisors. It is necessary for information about the occlusion, orthognatic or disgnatic (after Krunkenberg).

3.2.1.3. Determining gender by forensic-dental methods

(Zuhrt et al. 1978; Gustafson 1966; Lunth & Lunth 1973)

In the study, it was possible to use limited options since only 4 teeth were preserved.

The index of the series of dental rows after Pont – which provides information on the shape of the hard palate: this is more ellipsoid in females, more rounded in males. Pont was the first to devise the table of average values of the dental row.

- P = (width* x 100) / 80   * width between the first premolars
- M = (width** x 100) / 60   ** width between the first molars

The width of the dental row is in proportion to the width of the incisors. The width between the first premolars left and right is one fourth larger, and between the first molars left and right one third larger than the sum of the widths of the upper four incisors. The points of measurement are the deepest distal points of the first premolars and the deepest points of the frontal fissures of the first molars. They are measured by means of the orthodontic symmetroscope after Schwartz. There exists a correlation: one half of the distance between the zygomatic arches is the greatest width of the dental row.

3.2.2.
Use of Identification Methods for Determining Age by Skull

According to the morphological features of the skull, anthropologists divide the human lifespan into five periods (Hunger, Rother 1978; Ferembach 1979; Stewart 1954):

- infantia: ending between the ages of 12 and 14;
- juvenilitas: from the emergence of the second premolar to the end of growth and physical development. Concluding between the ages of 20 and 25;
- adultus or virilitas: a stable period, when the skeleton does not change, ending in males around the age of 45, in females around 40 years;
- maturitas: lasting up to the age of 55, when the processes of senility begin to occur;
- senium: accompanied by involutionary processes.

Nemeskéri divides man's lifespan into the following periods:

- infans I. (0–6 years)
- infans II. (6–13)
- juvenilis (13–20)
- adultus I. (20–40)
- adultus II. (40–60)
- maturus (50–70)
- senilis (over 70)

In general, the female skeleton ages faster than the male; paradoxically, however, women on average may expect a longer lifespan (Hunger, Leopold 1970). The oldest method for determining age by skull is based on the obliteration of the skull sutures. This method is not wholly reliable (Powers 1962), since it also occurs in physiological cases that the sutures do not close (Perizonius 1984), despite advanced age (Hunger, Rother 1978; Kemkes-Grottenhaler 1996; Krogman 1987). Thus, for instance, the **medial palatine suture obliterates late or not at all.**

Following the experience of anthropologists, it is possible to determine age between the ages of 20 and 45 years with an accuracy of 5 years, while after 60 the range of accuracy is 10 years.

Additional data on age are given (on the skull) by the **deep arterial grooves of the medial meningeal artery** on the parietal bone of the skull, and by the **more numerous, Pacchionian hollows**. Usually, they indicate an age of over 60. The muscle attachments are also more pronounced after the age of 25 (Hunger, Leopold 1970; Dominok 1959).

Since the cranial sutures provided the author with almost the only data on age – there were, in fact, no long bones – they were thoroughly studied. In this study, among 18 skulls – which represent a small sample (not suitable for statistical processing) – uniform criteria were adopted.

Numerous researchers have compared the differences in obliteration of ecto- and endocranial skull sutures in males and females. Ascádi and Nemeskéri have written that the processes do not differ (Ascádi, Nemeskéri 1970), yet more recent studies have established that obliteration in males occurs earlier in life, while in females it takes place later but faster (Kemkes-Grottenhaler 1996). Ossification **begins on the endocranial sutures** and continues on the ectocranial. Among the endocranial sutures the first to be obliterated is the coronal, followed by the sagittal and lambdoidal. In the ectocranial sutures the process is the reverse: first the sagittal suture closes, followed by the lambdoidal, while the latest to close is the coronal. The description above indicates the importance of observing endocranial and ectocranial obliteration in determining age by skull (Ascádi, Nemeskéri 1970).

Ascádi and Nemeskéri further built upon Martin's system for assessing the ossification of the skull sutures. They divided the coronal and lambdoidal sutures each into three parts: Cl, C2, C3 and Ll, L2, L3, and the sagittal into four parts – Sl, S2, S3 and S4. The course of ossification of the sutures was divided into four stages:

- 0 – suture is open
- 1 – beginning of closing
- 2 – closing is in progress
- 3 – closing has advanced
- 4 – suture is closed

The stages of obliteration of the sutures are linked with the supposed age (picts. 8, 9). Various authors have reported on the relative unreliability of this method for determining age (Powers 1962; Brothwell 1972; Key 1994).

On account of the small number of samples for statistical processing in the present study, objective results could not be given.

3.2.2.1. Determining age by obliteration of the endocranial sutures

(Hunger, Rother 1978)

- **sagittal suture:**
  pars obelica – begins between the ages of 15 and 17, ends at 35,
  pars bregmatica – begins between the ages of 23 and 27, ends at 43,
  pars verticis – begins between the ages of 20 and 22, ends at 45,
  pars postica or pars lambdica – begins between the ages of 26 and 28, ends at 46;
- **coronal suture:**
  pars bregmatica – begins between the ages of 22 and 28, ends at 49,
  pars complicata – begins between the ages of 19 and 23, ends at 49,
  pars temporalis or pars pterica – begins between the ages of 23 and 27, ends at 56;
- **lambdoid suture:**
  pars lambdica – begins between the ages of 30 and 35, ends between 59 to 65,
  pars media – begins between the ages of 32 and 35, ends at 56,
  pars asterica – begins between the ages of 40 and 43, ends at 56.

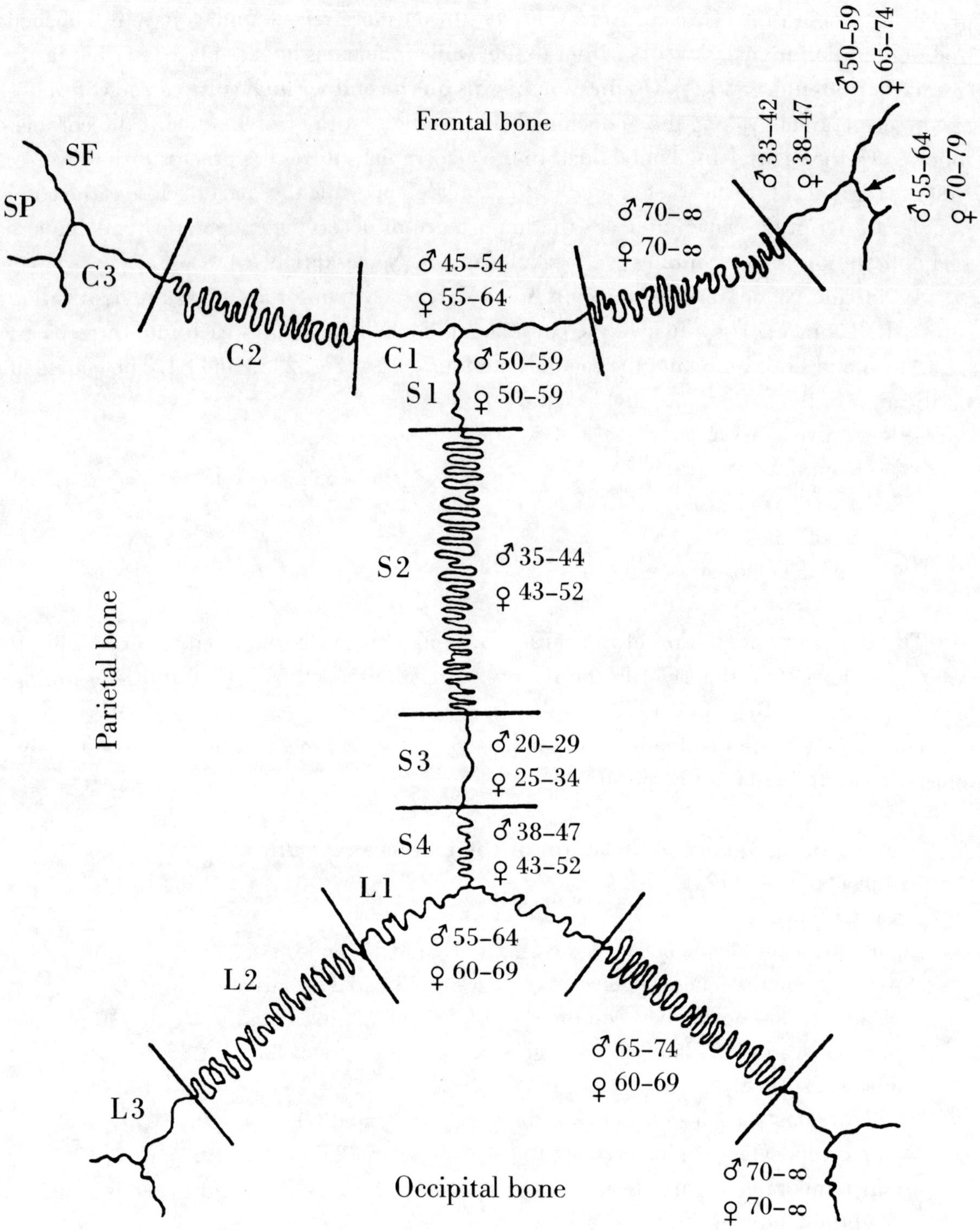

Picture 8: *Determining age by obliteration of the skull sutures I. (after Vallois 1937, modified by Rösing 1977)*
Legend: *C – coronal suture, S – sagittal suture, L – lambdoid suture*

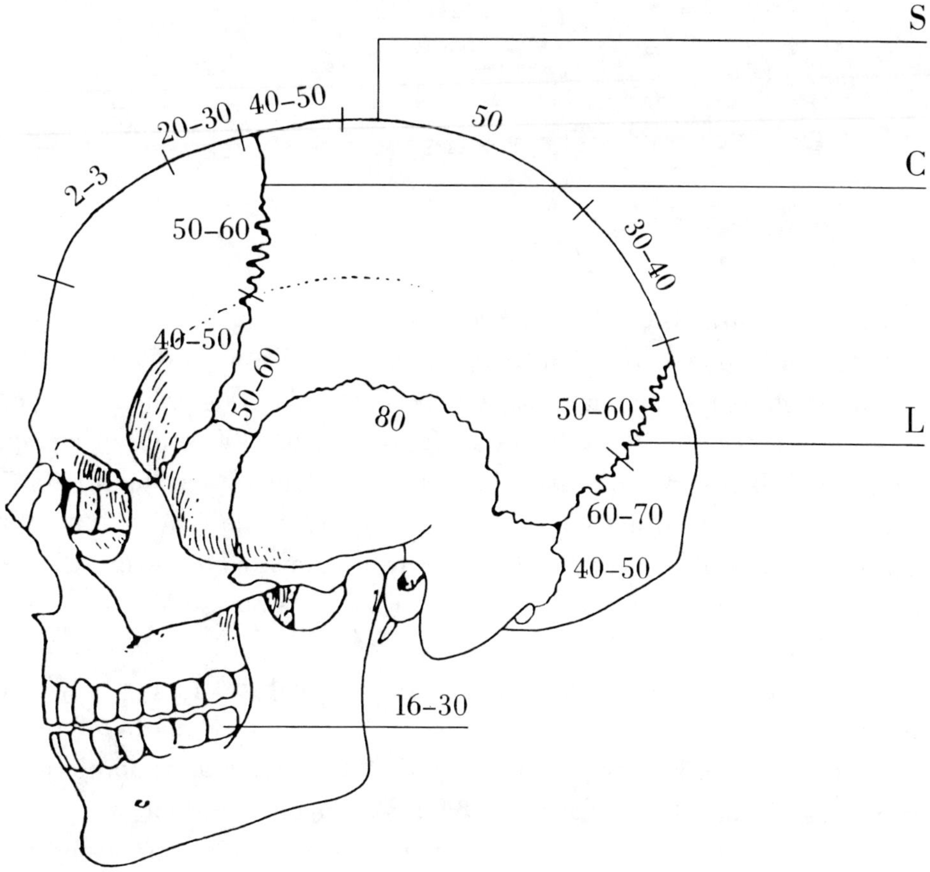

Picture 9: *Determining age by obliteration of the skull sutures II*
Legend: *C – coronal suture, S – sagittal suture, L – lambdoid suture*

3.2.2.2. Determining age considering obliteration of the ectocranial sutures

(Hunger, Rother 1978; Kemkes-Grottenthaler 1996)

– **sagittal suture:**

pars obelica – begins between ages 16 and 18, ends at 36,
pars bregmatica – begins between ages 26 and 29, ends at 46,
pars verticis – begins between ages 16 and 19, ends at 46,
pars postica – begins between ages 23 and 28, ends at 43;

– **coronal suture:**

pars bregmatica – begins between ages 22 and 26, ends at 60,
pars complicata – begins at age 21, ends at age 65,
pars temporalis – begins between ages 32 and 38, ends at 68;

– **lambdoid suture:**

pars lambdica – begins between ages 36 and 41, ends at 68,
pars media – begins between ages 38 and 43, ends at 69,
pars asterica – begins between ages 43 and 48, ends at 68;

3.2.2.3. Determining age by X-ray methods

(Hunger, Leopold, Wunderlich 1978; Grashey, Birkner 1964; Loepp 1954)

Although X-ray examination is equally valid to dactylographic, both being highly precise – Sassouni (Sassouni 1957), for example, proved on 500 skull radiographs that with five characters it is possible to reliably identify an individual – in the present instance it was of less assistance. Radiographs of the head are useful in identification if one has access to the appearance of the deceased; the identity may be compared by means of superprojection. That method could not come into consideration in this study, since no images of the deceased were available.

Determining age by X-ray methodology is a good instrument only with skeletons up to the age of 20, because the bone nuclei are still visible and can therefore be used in order to determine the age. For the middle-age period, the radiographs do not provide such reliable data on age (Tabor, Jevtič, Pavčnik 1990). For persons under 20 years of age:

- classical radiographs are used, in six projections: anteroposterior (fronto-occipital), postero-anterior (occipito-frontal), both lateral and both basal views;
- if necessary, computer tomograms are made (CT);
- all preserved teeth are X-rayed;
- orthopan-tomography is made of the maxilla.

3.2.2.4. Determining age by forensic-dental methods

(Zuhrt, Rottstock, Winterfeld 1978; Gustafson 1966; Lunth 1973)

Forensic medical knowledge is complemented by forensic dentistry. The firm biological structure of the teeth prevents their disintegration. Determining gender by teeth enters into consideration only if, on account of the decomposition or charring of the corpse, other methods of identification cannot be used. If 4 (four) of the front incisors are preserved, the age of the deceased may be assessed to a degree of accuracy of ± 6 years. This does not hold true for the premolars and molars (Zuhrt, Rottstock, Winterfeld 1978).

Such dental determination is of great assistance up till adolescence or adulthood, when the third molar is usually not yet developed; however, this tooth also does not provide us with reliable information about age, since it grows out over a lengthy time span (between the ages of 16 and 30). The exchange of milk teeth for permanent ones begins after the age of six, following the natural principles described in the literature. In children, their age can be estimated according to the stage of growth of the permanent teeth (Bikar 1970):

Development of permanent teeth:
- between ages 6 and 8, first molars,
- between ages 6 and 9, first incisors,
- between ages 7 and 10, second incisors,
- between ages 9 and 13, first premolars,
- between ages 11 and 14, second premolars,
- between ages 10 and 14, second molars,
- between ages 16 and 30, third molars.

In determining the age of teeth one observes (Zuhrt, Rottstock, Winterfeld 1978):
- tooth abrasion: this method had proved to be successful throughout past history; today, however, with the fundamental changes in human nutrition, its significance has been lost;
- change in tooth coloration: teeth of the younger are usually lighter, those of the older are darker; this method, likewise, is today no longer reliable;
- histological and biochemical changes;
- construction of the secondary dentine;
- transparency of the dental root.

In the research presented here, these methods are not applicable, since only four teeth altogether were preserved.

The width of tooth as a specific gender feature has not yet been dealt with. Also the occlusion is gender specific in 58% of cases, nondescriptive in 24%, while in 18% it even bears the features of the opposite sex.

No correlation exists between the craniometric and odontometric measurements (Anderson 1973).

3.2.2.5. Determining age by radioactive carbon ^{14}C

The ^{14}C radioactive carbon test is frequently used in determining the age of very old bone remains. On account of its reliability, this method is fundamental, and makes it possible to determine approximately to which period a skeleton belongs. The variability factor with this test is ± 30 years. For the present research, this method could not be used, since it would only have allowed one to establish whether or not the skulls actually did date from the 14th or 15th century.

3.2.3.

Use of Identification Methods for Determining Cause of Illness by Skull

3.2.3.1. Paleopathological methods

Paleopathology is a science which is concerned with studying illnesses in prehistoric times (Grmek 1989; Moodie 1923; Brothwell 1972; Steinbock 1976; Ortner, Putschar 1985; Katolyi 1970). In instances when, in addition to bones, other tissues are preserved – e.g. in mummies – these provide exhaustive data on the development of disease throughout time and on diseases of the past. If we have the possibility to study only skeletons, this is referred to as osteoarchaeology.

In paleopathology the borderline between the healthy and the sick is frequently blurred, and consequently, for examination, in addition to macroscopic methods it is also necessary to use microscopic methods (paleohistology, paleobacteriology, paleoimmunology, microroentgenology, electronic microscopy, etc.). Thus reliable findings are obtained.

On old bone remains, the following can be revealed (Lovrinčević, Mikić 1989):

– injuries (if the fracture has healed, one can see the callus) and fatal injuries to the skull, and trepanations on skulls. Today, in cancer diseases, metastases occur in the bones in approximately one quarter of cases. On ancient bones, this is almost never found;

– avitaminosis: on the bones only the traces of **avitaminosis** D and C are left. The former as **rickets** in children, or osteomalacia in adults, and the latter as **scurvy**. Rickets in children is manifested by deformations of the bones as a result of their insufficient mineralization. On the skulls of such children one observes a thickened occipital bone. Paleopathological studies have not observed scurvy up till the Middle Ages;

– rheumatic diseases: these rank among the most frequent osteoarchaeological diagnoses. They include: degenerative changes, changes caused by infections, digestive disorders, autoimmune reactions, inherited diseases and other causes. On old bones one frequently detects arthrosis with osteophytes, osteoporosis, wearing down of the temporomandibular articulating surfaces and tooth abrasion. Arthritis is difficult to demonstrate aetiologically (Grmek 1989). Infections rarely leave traces on the skeleton. Fistules do appear after osteomyelitis, and changes after treponematoses, leprosy and tuberculosis. Leprosy leaves exulcerations and mutilations on the bones; they most frequently occur on the facial bones, and can be revealed by X-rays. Among the osteoarchaeological finds there are few leprous changes (Grmek 1989). Tubercular changes on the bones most frequently appear on the spine and long bones, but rarely on the skull. In the Mediterranean, tuberculosis was already appearing from the Early Stone Age onwards (Grmek 1989). Viroses on the bones do not leave changes, with the exception of poliomyelitis. The same holds for fungal diseases.

In this research, macroscopic methods were used, and with them the pathology was verified, hence pathology as consequences of:

- bone injuries;
- avitaminosis: rickets, osteomalacia, scurvy;
- haematological abnormalities, visible on the skull: thalasemia, sideropenic anaemia, etc. Since among the skulls analysed there was one with a change on the ocular arch – cribra orbitalia – especial attention was paid to the relevant literature (Moseley 1961; Hengen 1971; Bergman 1993; Schutkowski, Grupe 1997; Aufderheide, Rodrigez-Martin 1998);
- rheumatic diseases and degenerative changes: arthrosis, wearing down of the temporomandibular articulating surfaces, tooth abrasion, osteoporosis, etc.;
- non-specific inflammations: osteoarthritis, etc. (Panders, Hadders 1970);
- specific inflammations: treponematosis, leprosy, tuberculosis;
- digestive disorders;
- inherited diseases;
- tumors: primary bone tumors, metastases of tumors into the bone.

Since, after the preliminary part of this research, it was expected that the pathology on the maxilla of skull 14 would be linked with gigantism, the author examined the related literature, consulted with pathologists, and differentially diagnostically restricted herself to the following diseases:

- ossifying fibroma: a benign bone tumor with slow growth, which can appear anywhere on the bones, also in the maxilla, and may greatly alter the bone.
- **fibrous dysplasia: an idiopathic disease**, which normally affects the marrowbones and replaces the tissue with poorly organized fibrositic tissue. It grows slowly and progressively, rarely aggressively. Deformities appear. If they are on the face, they lead to a **monstrous appearance, pains**, altered functions and **pathological fractures**. In older patients the bone changes are more sclerotic. Radiotherapy does not affect the growth of fibrodysplasia; the **therapy of choice is surgery**. Fibrous dysplasia rarely undergoes malign alteration; according to the statistics only in 1% of cases. If, however, it does turn into a malign sarcoma, the average period of this alteration is about thirteen and a half years. In this instance the prognosis is poor. Most patients die as a result of metastatic lung tumors. The survival period is approximately 3 years. The disease does not show up well on radiographs.

Fibrous dysplasia occurs in three forms: **monoostotic**, **polyostotic**, and as **Albright's syndrome**. In the first instance, the most frequent location is precisely on the maxilla, then on the mandible. On the maxilla, it is the zygomatic process above the first molars that is most frequently affected. Polyostotic fibrous dysplasia represents approximately one quarter of all fibrous dysplasias. In almost half of cases the cranial base is affected.

Albright's syndrome or the McLune-Albright syndrome accounts for **1–3% of fibrous dysplasias** and occurs more frequently in males. It is accompanied by skin pigment marks, which are more frequent on places above the bone lesion. Darker coloration may also occur on the mucosa of the lips, the hard and soft palate and the buccal mucous membrane. In certain cases, premature sexual development is described, with the early appearance of the menarche in girls. It seems that pubertas praecox is a rare occurrence in males. The aetiology of this accompanying occurrence is not known. Formerly, the opinion was held that, in cases in which **fibrous dysplasia affects the cranial base**, this may cause **aberrations in the axis of the hypothalamus-hypophysis-sexual gland.**

However, cases have also been described in which this axis was disturbed, yet premature sexual maturity did not occur. Contemporary, highly sensitive methods – such as bioassay and radioimmunoassay – have shown that sexual or other **endocrinal hyperfunctions** are the consequence of the **autonomous activity of the terminal organ**, and not of the influence of the central nervous system. Recently, in Albright's syndrome, **mutation** has been discovered in the **gene** which encodes the **sub-unit of the stimulatory G protein adenilate cyclase, and this was recognised as the cause of excessive endocrinal activity in Albright's syndrome.** In conjunction with this the following were noted: nanosomia, goitre with hyperthyroidism or without, hyperparathyroidism, diabetes, Cushing's disease, acromegaly, coarctation of the aorta, heightened blood pressure, rudimentary kidneys, and mental retardation (Barnes 2001).

– Osteosarcoma

If one excludes malignant melanoma, osteosarcoma is the most frequent bone tumor, representing approximately 19% of all primary malignant neoplasms. It is almost twice as frequent as chondrosarcoma, three times more frequent than Ewing's sarcoma, and ten times more common than malign fibrous bone hystiosarcomas. Osteosarcoma generally occurs in young persons at the end of the growing period and is most often detected around the age of 25. It can, however, also occur in adults. Usually the long bones are affected, particularly on the distal parts of the thighbone or the proximal parts of the tibia, hence around the knee. In less than 10% of cases bones of the head are affected, most frequently the lower and upper jaw. Approximately 6.5% of all osteosarcomas occur at this place and are somewhat more frequent in males, on average around the age of 36. The maxilla is more rarely affected, the alveolar arch being the part most often concerned. This is accompanied by the loss of teeth, metastatizing most frequently in the lungs, in approximately 1/3 of cases, with an average survival period of 5 years. Radiographically, osteosarcoma presents a diverse picture: destructive, poorly organized, with central bone lesion, etc. The lesion may be lithic, sclerotic or mixed. Among the tumors included in the differential diagnostic picture are chondrosarcomas, fibrous dysplasias, and similar (Barnes 2001).

Potential diagnoses also include:
- gigantocellular granulomas
- eosinophilic granulomas

3.2.4.
Use of Identification Methods for Determining Family Linkage between Skulls

3.2.4.1.
Genetic Methods

In order to determine family linkage, genetic analysis is the method of choice.

There are very few laboratories worldwide which are engaged in the genetic research into paleopathological and archaeological material. Consequently, the information for this study had to be sought in the appropriate literature and in collaboration with reference laboratories worldwide.

During the last fifteen years, in molecular genetics, methods of DNA research have been developed with the aid of which it is possible to identify persons with exceptional precision and to authenticate kinship links between individuals. The success of these methods depends on the quantity and quality of the isolated DNA, i.e. on the quality of the biological materials which we wish to research. **For fresh tissues and well preserved biological materials, the genotyping methods are the most successful.** However, very old organic materials (hundreds, even thousands of years old) enable us to isolate exceptionally small quantities of DNA, which is frequently badly damaged, and this consequently has considerable influence on the success of the genetic analyses. A notable example of the usefulness of the genetic methods of analysis for human remains several decades old is the identification of the remains of Josef Mengele (Jeffreys 1992). Since his biological material was not badly damaged, they succeeded in isolating even nuclear DNA. **In very damaged samples, it is possible at most to isolate the mitochondrial DNA.** In the case of the identification of the Russian imperial family, the Romanovs, the procedure did indeed involve **mitochondrial DNA (Gill 1994).**

The successful identification of the skeletal remains of the Romanov family made us reflect on the possibility of identifying the remains of the skulls studied by means of DNA analysis. This way, the answers recorded in the study would be more specific and perhaps even more final. In the case of the Romanov family, the analyses were carried out on the nuclear and mitochondrial DNA. With the aid of the nuclear DNA they verified the kinship links between the supposed members of the family (relationship of family members after mother and father), however they were not able to answer the question of whether these were indeed the posthumous remains of the Romanov family. The **nuclear DNA actually permits only the verification of closer kinship relations (between children and parents), however, on account of recombination, it is not possible to verify kinship between distant relatives who are separated from one another by several generations.** The only way of verifying **distant kinship relations** is by **analysing the mitochondrial DNA**, since it is only this that is **inherited exclusively through the mother and does not recombine** (Lutz 1996). In the Romanov family, they verified the identicality of the skeletal remains by comparing the mitrochondrial DNA with that of still living descendents on the maternal line several generations apart. **By analysis of the nuclear DNA they determined the gender** for all the skeletal remains of the Romanov family (Gill 1994). For this purpose, in the **polymerise chain reaction (PCR) they multiplied a very short marker (in length approximately 100 basic pairs)**, which permits determination of gender even in partially damaged DNA (Sullivan 1993).

From the Counts of Celje, there have been preserved only the skulls, which are presumed to have belonged to them, and by comparison with the skeletal remains of the Romanov family they are far older (supposedly 500–600 years); consequently, the success of isolating the nuclear DNA from them was highly questionable, since the probability of splitting was so great that DNA analysis might not be possible. It might, indeed, be possible that there was no longer any DNA left in the skulls, or else that it was so badly damaged that it could not be analysed. Since gender is determined exclusively by means of the nuclear DNA markers, and since it was anticipated that in such old bones the nuclear DNA would be severely damaged, this analysis probably would not be possible (gender remains questionable in two of the children` s skulls).

Geneaological relationship between the skulls could be established only through **mitochondrial DNA**, since the latter – in contrast to nuclear DNA – is found in the **cell with a great number of copies** (Holland 1993, Hagelberg 1992) and is **preserved longer (it is protected by its conformation against decomposition by exonucleases – the circular molecule – and the mitochondrial membrane)** (Hopwood 1996). Nevertheless, the analysis of the mitochondrial DNA would reveal only the **maternal relationship of the presumed family members**, but there would not be obtained any information on potential relationship through the paternal line. The genetic research on the skulls which are presumed to have belonged to the Counts of Celje became especially interesting when it was discovered that there were still living blood descendents from the line of Barbara of Celje.

The findings, provided by performable genetic research, would be derived from verification of the relationship between the supposed family members on the **maternal line with the aid of mitochondrial DNA**. Considering that persons with a common female ancestor must have the **same nucleotide sequence** of mitochondrial DNA, it may be predicted that from the genealogical line of the Counts of Celje there would be 8 different sequences of mitochondrial DNA, of which 5 sequences (2, 3, 4, 5, 7) would be repeated in more than one of the skulls; this would be proof that they had a common female ancestor, i.e. that their heredity was on the maternal line. Hence, in the analysis of 18 skulls, one expects to find 8 different sequences of mitochondrial DNA:

- sequence 1: Friedrich I
- sequence 2: Didmund of Walsee – Ulrich I, Hermann I of Celje
- sequence 3: Adelaide of Ortenburg – Wilhelm
- sequence 4: Catherine Kotromanić – Hans
- sequence 5: Anna of Schaunberg – Friedrich II of Celje, Hermann III of Celje, Ludwig
- sequence 6: Elisabeth of Abensberg
- sequence 7: Elisabeth of Frangepan – Ulrich II of Celje
- sequence 8: Hermann IV of Celje

For test identification of the skulls, professional advice was sought from several institutions worldwide. Consultations were held with Dr. Peter D. Gill and the staff who had conducted the identification of the posthumous remains of the Russian imperial family of the Romanovs (Forensic Science Service, Birmingham, UK). They were prepared to undertake the research on the understanding that they could not determine the mitochondrial DNA sequencing for all 18 skulls. Considering that the skulls are more than 500 years old, it seemed likely to them indeed that there would be no DNA remaining in them, or that it would be split to such an extent that the desired analysis could not be undertaken. No other biological substances were available for study.

The second reference person was Prof. Dr. Svante Pääbo, with a group of experts who are working in Munich, and who identified the »ice man« found in the Tyrol, named Oetzi. Dr. Pääbo was of a similar opinion to Dr. Gill, that most probably the only feasible research would be on the mitochondrial DNA, which would demonstrate the relationship of the family members along the maternal line (Pääbo 1993). He expressed doubt concerning the feasibility of research on the nuclear DNA, which would be expected to yield information about relationship of the supposed family members along the paternal line. He, too, was interested in the study, but had to decline it on account of having too much work and too little time. A similar opinion concerning the research approach was held by Prof. Dr. Berndt Herrmann from Göttingen, the author of the reference book on mitochondrial DNA, Ancient DNA (Hermann, Hummel 1992).

Finally, success was achieved with the paleogenetic laboratory in Rome, who responded enthusiastically to the research. The work was accepted by Prof. Gian Franco De Stefano. Work on the samples was accompanied by numerous difficulties. The results showed **great contamination of the samples, and degradation**. It was not possible to isolate the mitochondrial DNA from them. In the research studies into the archaeological material, more than 80% of the analyses proved unsuccessful on account of contamination (Hermann, Hummel 1992). Nevertheless, this also was a result, even if negative.

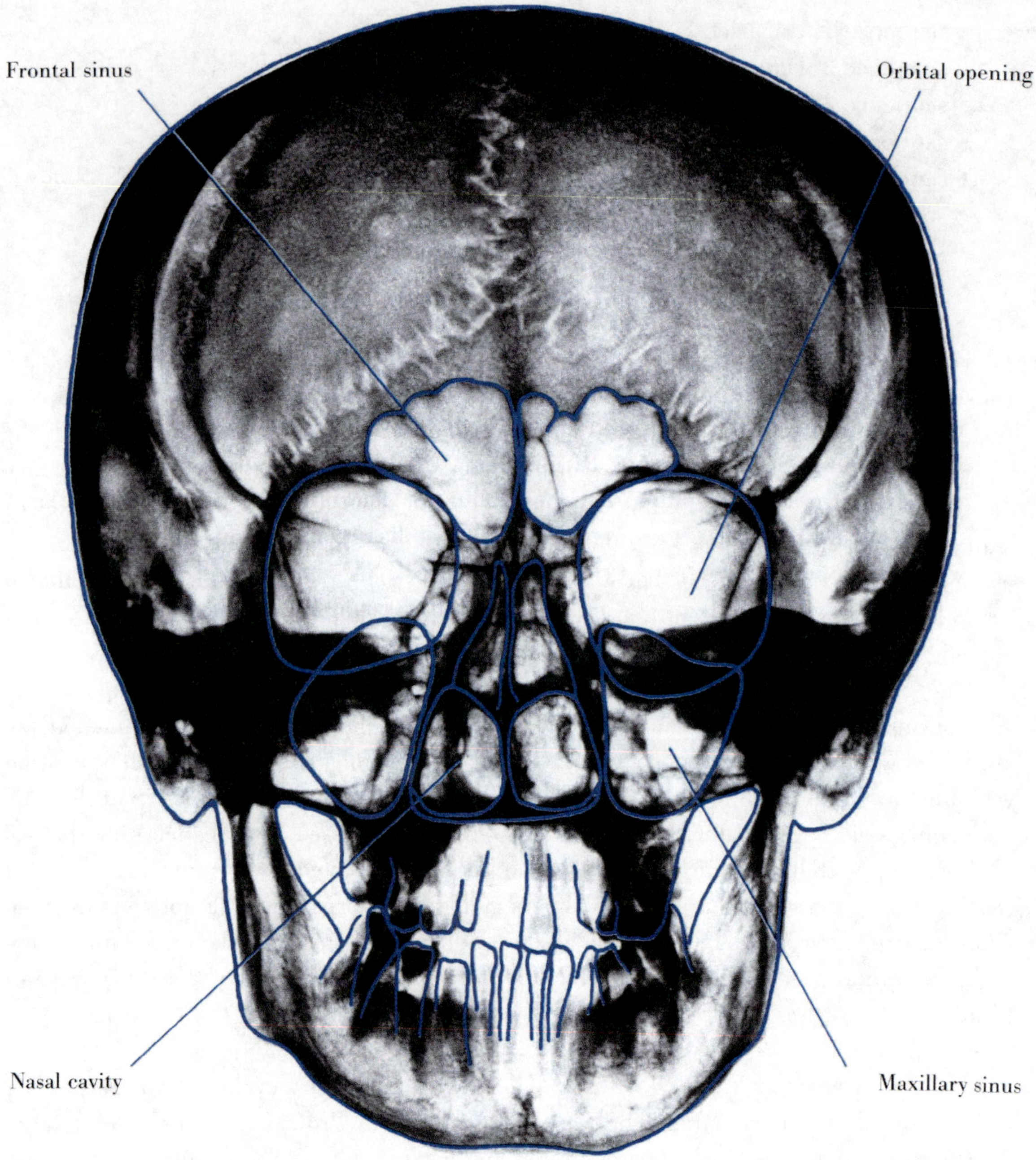

Picture 10: *Occipito-frontal radiograph showing the facial cavities (Hauser, De Stefano)*

3.2.4.2.
Epigenetic Methods

The modern »epigenetic identification method« is based on the morphological variation of minor skeletal traits observed on the human skull, and of their observed familial concentrations.

The term epigenetic was applied to these variants by Berry & Searle in 1963, and Berry and Berry (1967) defined epigenetic variants as an expression of genes affecting development. Synonyms which are or have been used for the term epigenetic variants are: minor skeletal variants, non metrical variants, quasi continuous, discontinuous, discreta. Hauser and De Stefano, the authors of the modern reference book on these traits consider the term epigenetic to be the most appropriate in view of the epigenetic control which, according to Hall (1984), modifies skeletal differentiation. Such control is exerted by factors, which are external to the responding cells and modify the environment in which the connective tissues develop or function. For each epigenetic trait studied, perception of these manifestations of complex genetic and environmental interactions requires knowledge of its nomenclature, gross anatomy, function, embryology and growth, genetics, medical relevance, sex difference, symmetry (if the trait is bilateral), laterality or side preference, age variation and associations with other traits.

Most important for the study of epigenetic traits is standardisation of the degree of expression, which is obtained by standardised methods of scoring. Thus several traits studied on the skull represent hyperostotic characters (Ossenberg 1970). **Hypostotic traits** indicate incomplete bone development and also retain the early bone developmental phases in adulthood. Hyperostotic traces indicate excessive bone growth. These do not necessarily appear only in adults, for they may also be visible in the earlier phases of bone development (Manzi, Vienna 1997). Certain apertures **(foramina) are the consequence of incomplete ossification** of the osseous lamina, while others may represent **hyperostotic processes.** Extreme prominence of certain traits may border on a pathological condition, which must be excluded in the research studies; otherwise, the **canals, grooves, protuberances** and other anatomical bone structures retain their function. They form the paths for the passage of the vessels and nerves, and serve as attachment points for the ligaments, tendons and muscles. The sutures form an important growth area along the bone margins. For certain structures, e.g. the **tori, their function has still not been elucidated.** Bone traits on the skull may also point to differentiation. **Verification of the epigenetic in relation to the genetic is not** always **straightforward**, since one is mainly dealing with bone material, and there are **few possibilities for researching identical twins**, and, in general, interpersonal family or kinship relations (Hauser, De Stefano 1989).

The knowledge hitherto of the epigenetic traits indicates little influence, or none whatsoever, of gender dimorphism. Likewise, the epigenetic traits in the period of adulthood are not dependent on age. They appear independently on one or the other side of the skull; nevertheless, one must always take into account the symmetry or laterality of these traits (Hauser, De Stefano 1989).

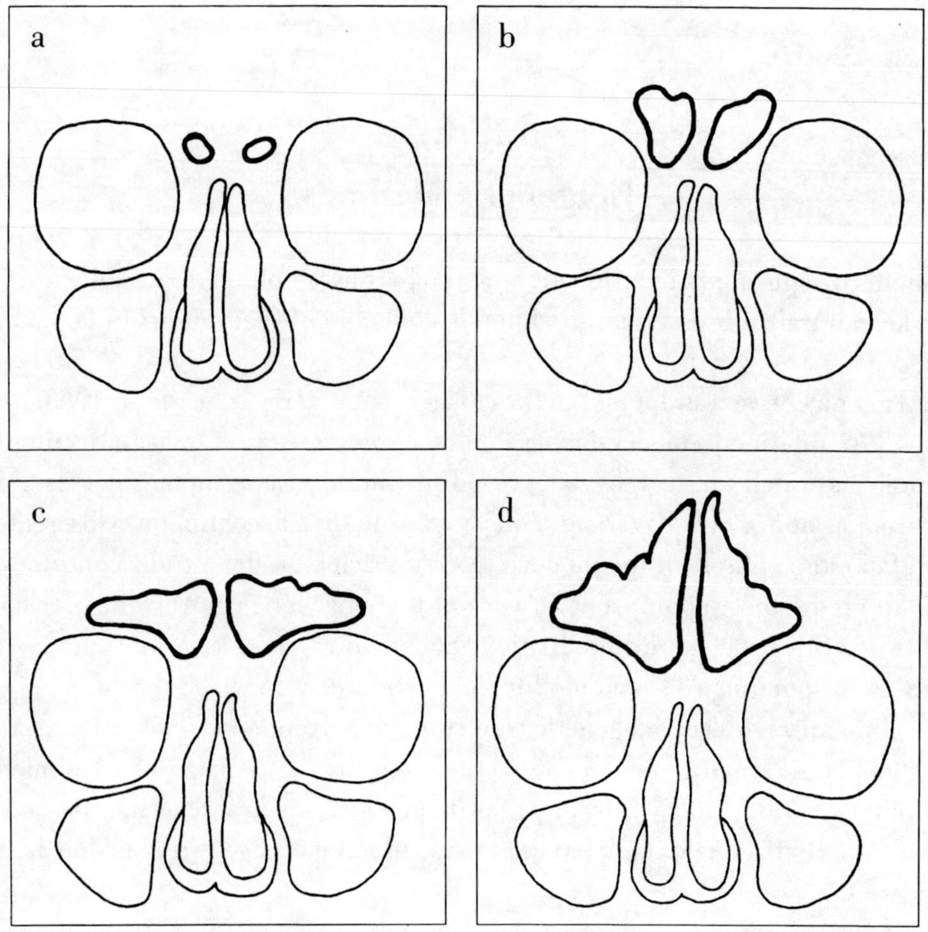

Picture 11: *The four categories of the frontal sinus: a) bean-shaped, b) leaf-shaped, c) fan-shaped, d) pyramid-shaped (according to Szilvássy 1986)*

3.2.4.2.1. Comparison of the facial cavities

In this research, the central method used was the comparison of the facial cavities of all 18 skulls. With regard to these, **X-ray studies** of even the oldest bones found reveal that the cavities have scarcely changed for over 100,000 years. The occipito-frontal radiographs provide a sketch of the frontal and maxillary sinuses, and the shape of the orbital openings and the nasal cavities (Szilvássy, Kritscher, Vlček 1987) (pict. 10).

Frontal sinus (pict. 11)

Leicher was first to draw attention to the strong genetic dependence of the shape and size of the frontal sinuses in 1928, and this was further demonstrated with more thorough research by Szilvássy on a large number of radiographs of family linkage between persons (Szilvássy 1986). **Following the older forensic medicine methods it was possible to use the shape** and size of the frontal sinuses as an aid in determining paternity. Szilvássy differentiated the frontal sinuses by shape into four categories: small and round, similar to a bean; somewhat larger, similar to a leaf; even larger, similar to a fan; and the largest, pyramidal in shape (pict. 11).

The shape of the frontal sinuses is determined by posteroanterior radiographs of the head. There are numerous variations in the size of the frontal sinuses in both sexes; asymmetry is almost the rule, though females usually have more symmetric frontal sinuses

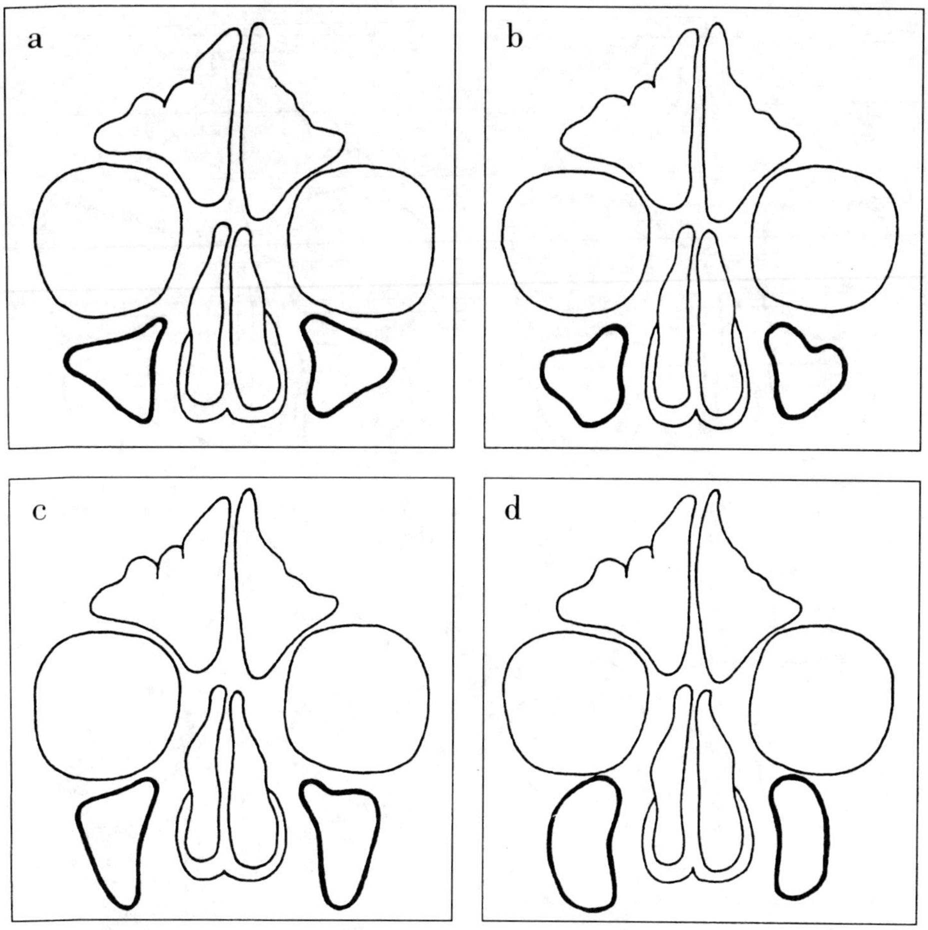

Picture 12: *The four categories of shape of the maxillary sinus: a) triangular, b) leaf-shaped, c) scapular, d) kidney-shaped (according to Szilvássy 1986)*

(Szilvássy 1982). The surface of the frontal sinuses is usually larger in males than in females. The pyramid shape is more frequent in females, while in males it is the leaf shape. In both sexes, the small bean shaped frontal sinuses occur very rarely, most frequently they are **fan shaped.** The left sinus usually dominates in size over the right. In brachycephalics the frontal sinuses are less developed than in dolicocephalics (Szilvássy 1982). Since this research included also children's skulls, the findings on the gender difference in the growth and development of the frontal sinuses are of interest. Their **development begins with the first year of age**. Between the ages of 1 and 5, the frontal sinuses develop similarly in girls and boys, and **around the age of 6 attain the size of a pea**. Between the ages of 8 and 12, the growth in boys is more delayed in comparison to girls, while between the ages of 14 and 15 their growth in boys overtakes that of girls. **Around the age of 18 the growth of the sinuses is completed** (Szilvássy 1981). The same author, together with colleagues in Austria, has carried out extensive research on the facial cavities in Zwölfaxing (Szilvássy 1974) and Stillfried, where seven kinship interlinked persons were buried in the family grave (Felgenhauer, Szilvássy, Kritscher, Hauser 1988).

Maxillary sinus (pict. 12)

By means of occipito-frontal radiographs, Szilvássy has demonstrated on a large number of samples of family interrelated skulls that above all the shape, though also the size, of the

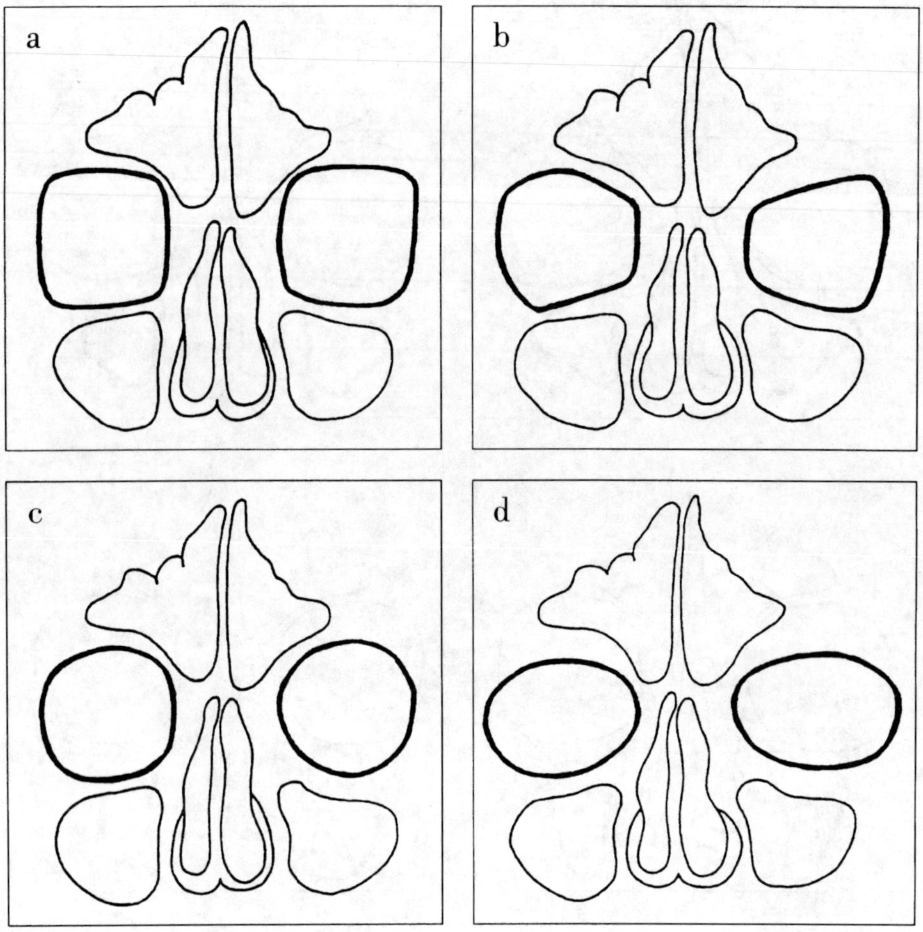

Picture 13: *The four categories of shape of the orbital opening: a) square, b) trapezoid, c) round, d) ellipsoid (according to Szilvássy 1986)*

sinuses of the maxilla are genealogically conditioned (Szilvássy 1986). He has introduced the methodology for evaluating this occurrence. He has differentiated four shapes of the maxillary sinuses into: triangular, leaf-shaped, scapular, and kidney-shaped (pict. 12).

The surface of the sinus in males is considerably larger than in females. In males it is on average 1.41 cm^2, in females 1.36 cm^2. In both sexes, the right sinus is usually larger than the left, and the triangular shape is more frequent than the leaf shape; the scapular and kidney shapes appear in only 17–21% of cases.

Orbital opening (pict. 13)

On a large number of samples of genealogically related persons, Szilvássy has demonstrated that there exists a strong genetic determination of the shape and size of the orbital openings, visible from the occipito-frontal radiographs of the head (Szilvássy 1986). According to shape, he differentiates four shapes of the orbital openings: square, trapezoid, round, and ellipsoid (pict. 13).

The surface of the orbital opening is considerably greater in males than in females (1.40 cm^2 and 1.27 cm^2, respectively). In humans (most of his research was carried out on Austrians), the most frequent shapes were square (54%) and trapezoid (33%), while the round and ellipsoid were rarer.

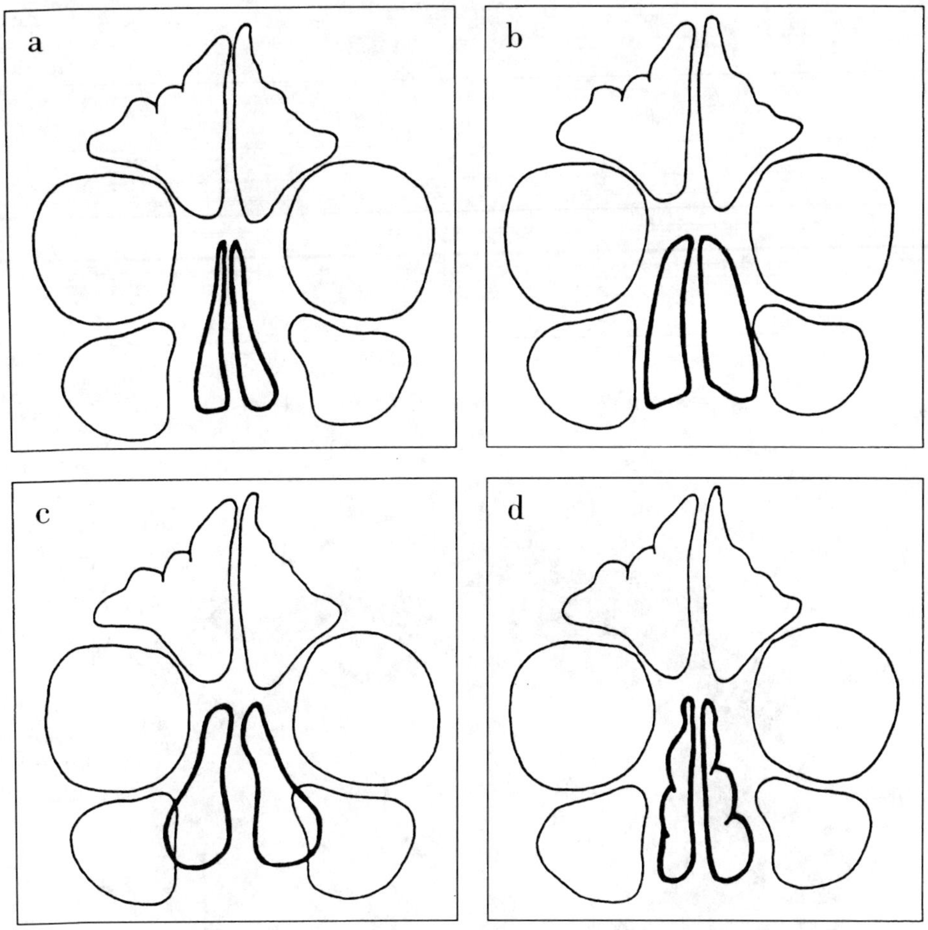

Picture 14: *The four categories of shape of the nasal cavities: a) triangular, b) trapezoid, c) pear-shaped, d) scalloped (according to Szilvássy 1986)*

Nasal cavities (pict. 14)

Likewise with regard to the shape and size of the nasal cavities, genetic determination is strongly present. The research was carried out by examination of numerous postero-anterior head X-rays of family members. Szilvássy differentiated four shapes of the nasal cavities: triangular, trapezoid, pearshape, and scalloped with indented margins (pict. 14).

The surface of the nasal cavity is considerably larger in males than in females (1.63 cm^2 and 1.07 cm^2, respectively). Asymmetry is frequent, and the left cavity is usually larger than the right. The shape of the nasal cavity reflects individual and population variations. The trapezoid and triangular shapes occur more rarely, while the pear-shaped and scalloped forms are found in altogether 41% of cases (Szilvássy 1986).

Analysis of the facial cavities on the skulls supposed to have belonged to the Counts of Celje yielded important findings.

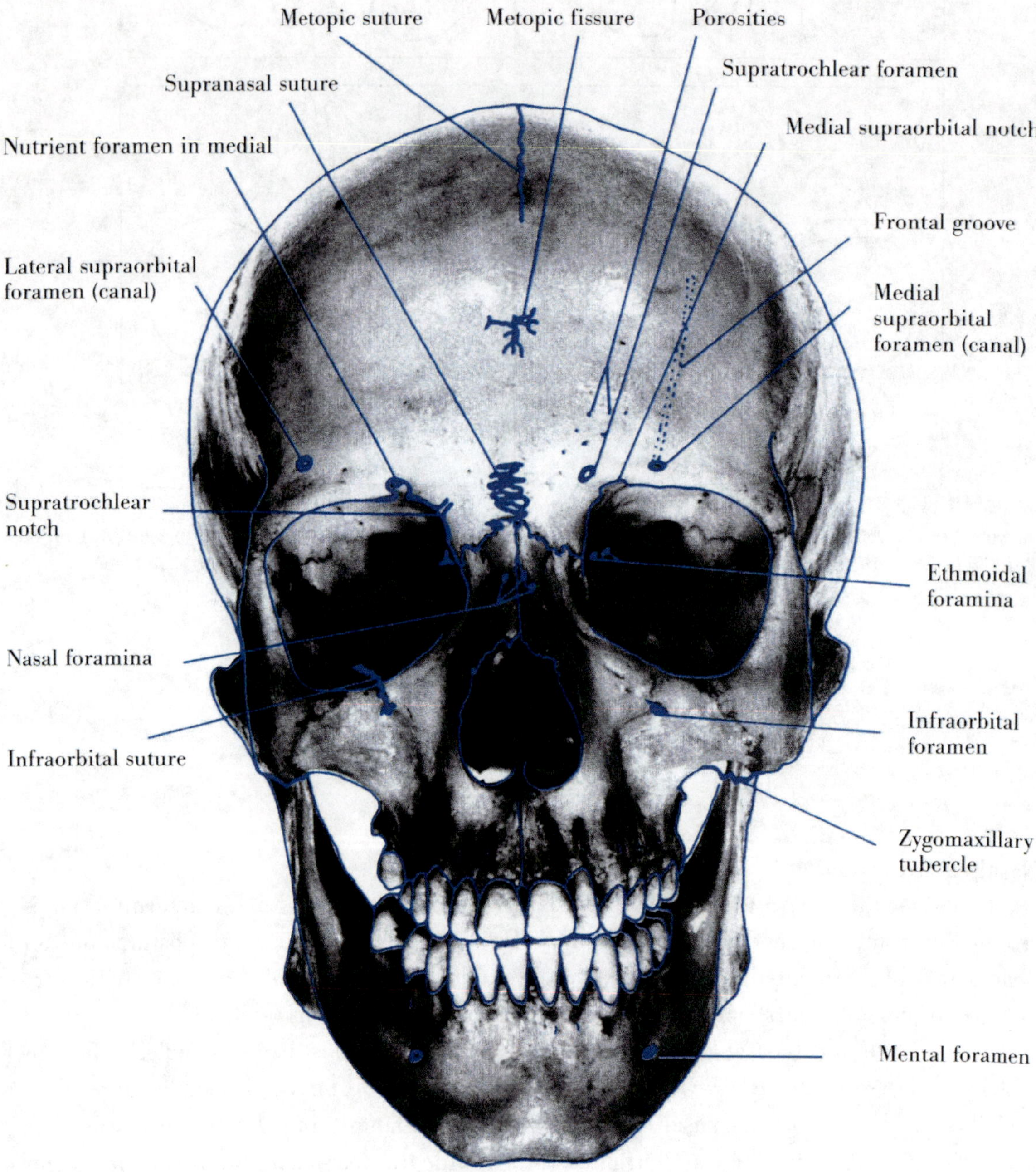

Picture 15: *Epigenetic traits on the frontal side of the skull (Hauser, De Stefano, 1989)*

3.2.4.2.2. Other epigenetic traits

Apart from the facial cavities (frontal and maxillary sinuses, orbital and nasal openings), an examination was made on each skull of approximately 30 of the most important epigenetic traits, and these were compared in terms of their presence or absence, greater or lesser prominence, symmetry, laterality, and other characteristic features described in the introduction to the chapter on epigenetics. Amongst the epigenetic traits, the studies carried out on the skulls involved mainly the following structures (picts. 15, 16, 17, 18, 19):

- supraorbital region:
  - supraorbital foramen (foramen supraorbitale)
  - medial supraorbital notch (incisura supraorbitalis medialis)
  - frontal groove (sulcus frontalis)
  - medial supraorbital foramen (canalis supraorbitalis medialis)
- supranasal suture (sutura supranasale) – infraorbital region
  - infraorbital foramen (foramen infraorbitale)
  - zygofacial foramen (foramen zygomatico-faciale)
  - nasal foramina (foramina nasalia)
  - infraorbital suture (sutura infraorbitalis)
  - zygomaxillary tubercle (tuberculum zygomaxillare)
  - marginal tubercle (tuberculum marginale)
- cranial vault
  - parietal foramina (foramina parietalia)
- mastoid region
  - mastoid foramen (foramen mastoideum)
  - exsutural mastoid foramen (foramen mastoideum exsuturalis)
- cranial base
  - hypoglossal canal (canalis hypoglossus)
  - basilar canal (canalis basilaris medianus)
  - spinous foramen (foramen spinosum dehiscens)
- palatine region
  - lesser palatine foramen (foramen palatinus minus)
  - transverse palatine suture (sutura palatina transversa)
- cranial sutures (picts. 17, 18, 19)
  - shape
  - course
  - particularities in the course (sutural ossicles: lambdoid, sagittal, coronal)
- additional ossicles of the skull (pict. 18, 19)
  - asterionic ossicle, inca bone and others.

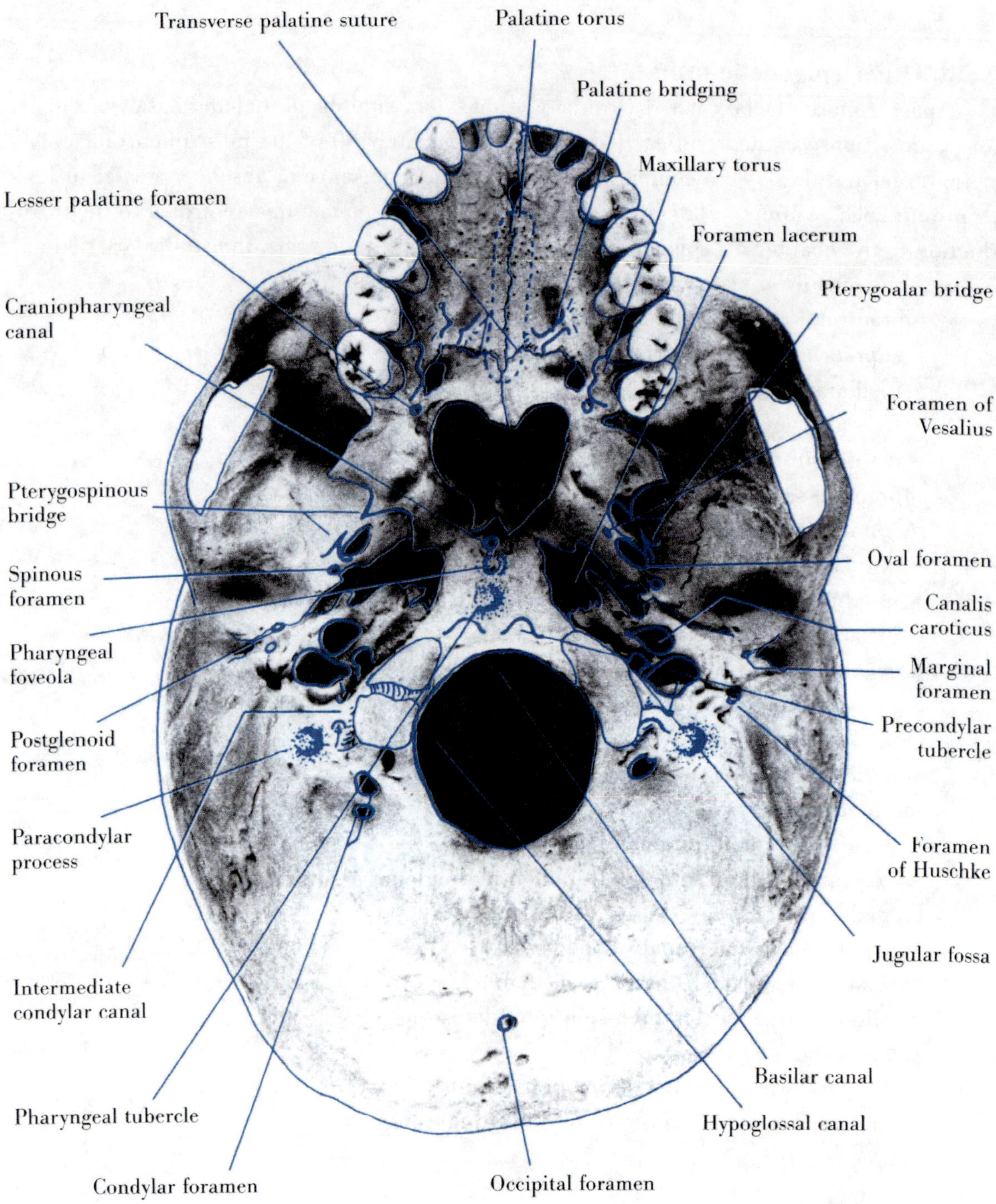

Picture 16: *Epigenetic traits at the base of the skull (Hauser, De Stefano, 1989)*

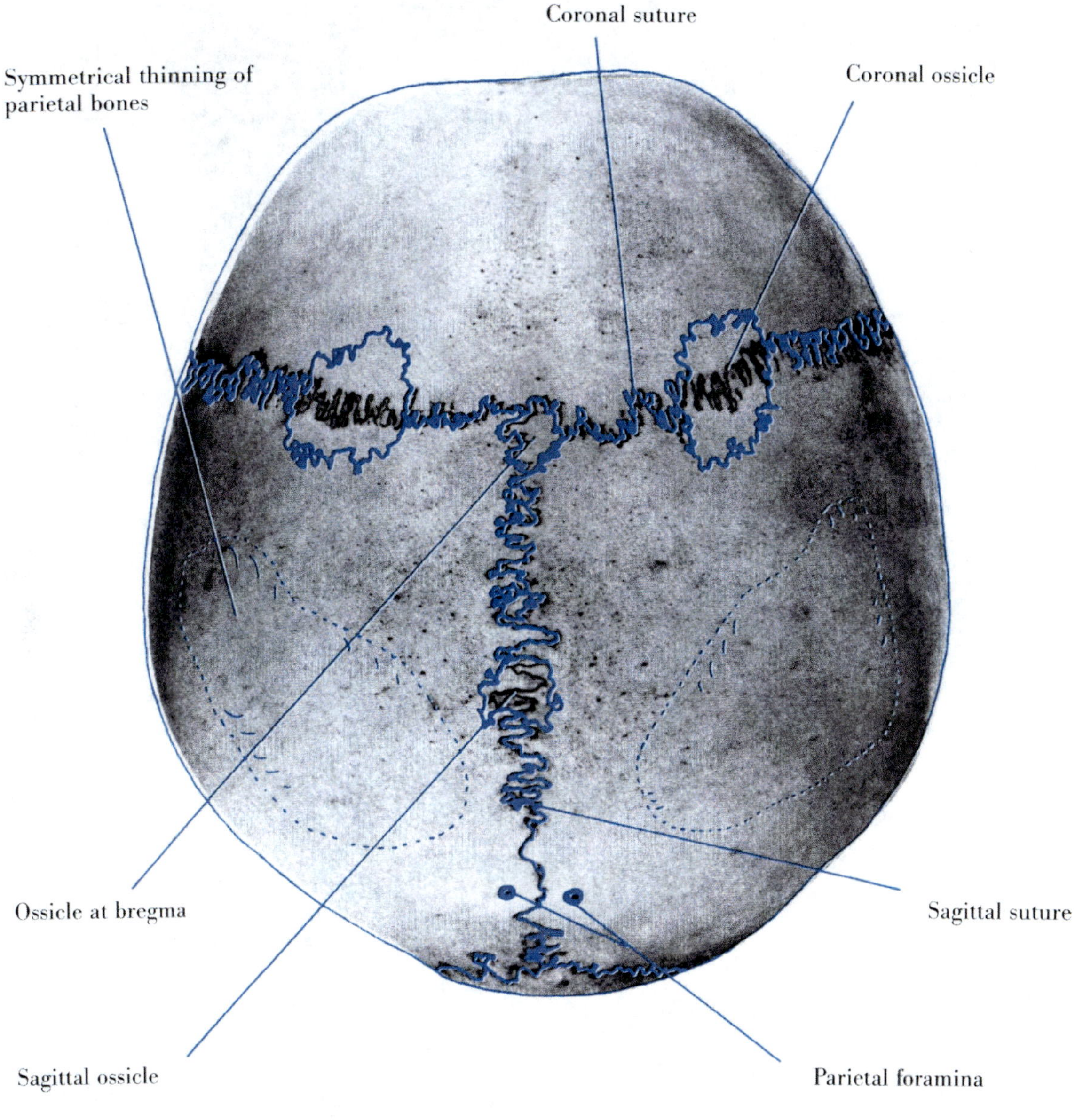

Picture 17: *Epigenetic traits and sutures on the cranial vault (Hauser, De Stefano, 1989)*

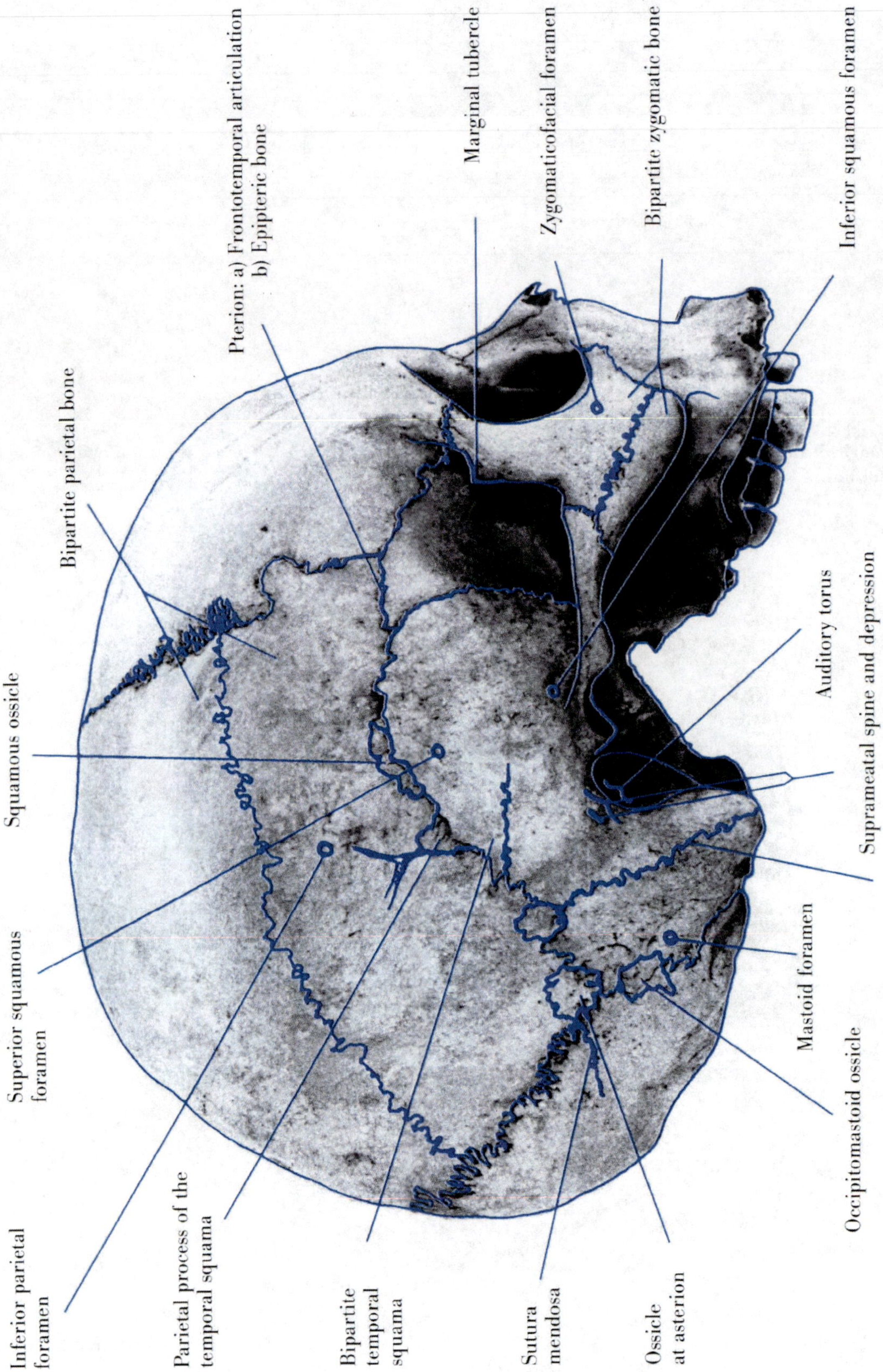

Picture 18: *Epigenetic traits and sutures on the temporal side of the skull (Hauser, De Stefano, 1989)*

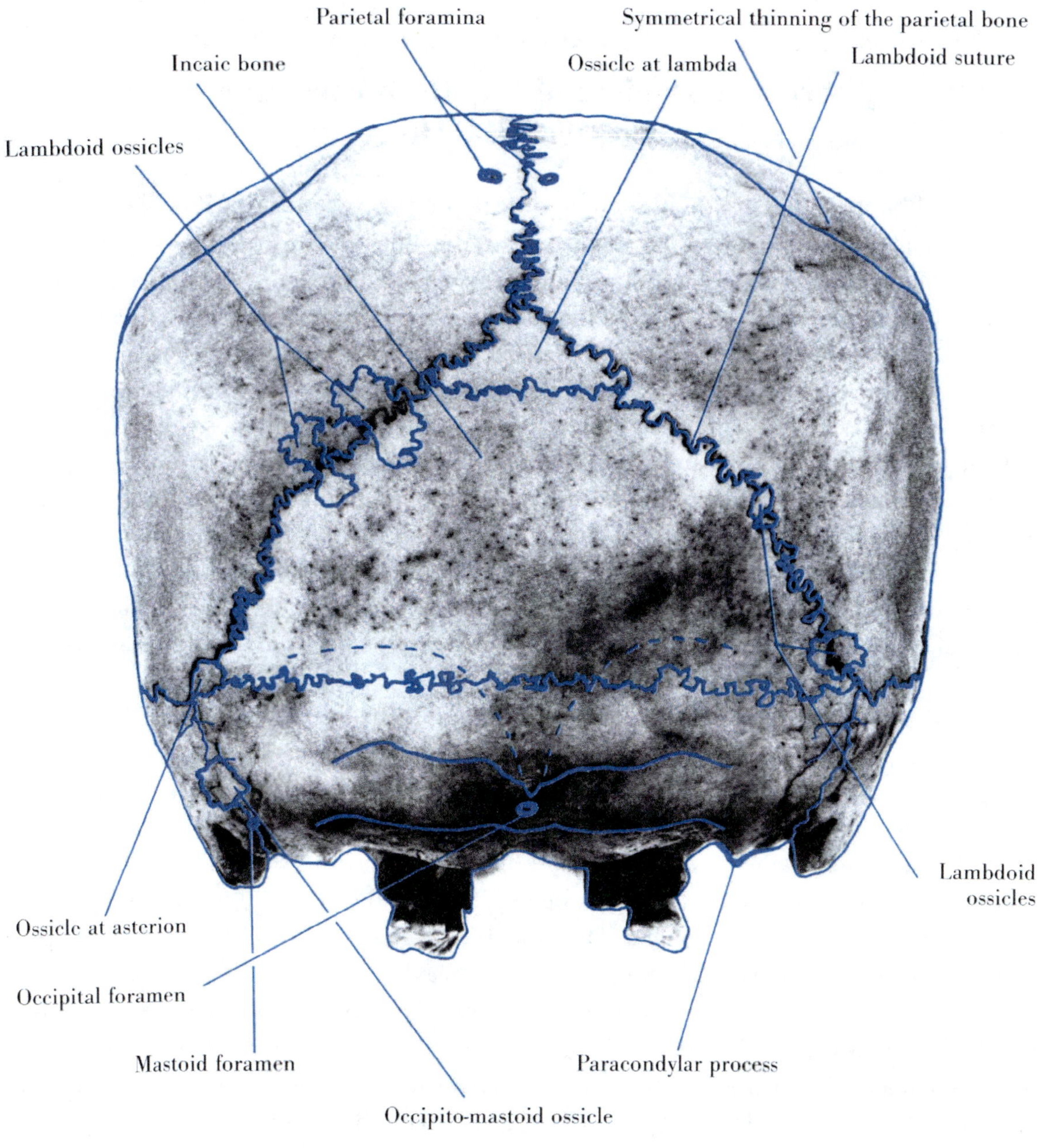

Picture 19: *Epigenetic traits and sutures on the cranial occiput (Hauser, De Stefano, 1989)*

3.2.5.
Art History Methods for Reconstructing Soft Facial Parts

The soft parts of the face are very different regarding age, gender, nutritional state, body structure and other influences. In reconstructing the soft parts of the face one must not underestimate their constitution: thus, muscular males have more pronounced muscle attachments on the skull, and a prominent bone microrelief; in them, there are also more soft facial tissues than in females (Leopold 1978, Boddington et al. 1987). There is no correlation between the size of the orbit and the eye-bulbs. The shape of the orbit, however, does determine the shape of the lower and upper eyelids. The shape of the nose-tip is decisively influenced by the shape, size and course of the spinae nasalis anterior. The course of the nasal margin and the position of the nose-tip are to a great extent dependent on the inclination of the nasal bone (Hunger, Leopold 1978). The schema of the nasal bones (pict. 20) determines whether the nose will be straight, irregular (upturned) or aquiline (Sailer 1959).

The pinnae, buccal cavity and lips have less influence on facial appearance. Posthumous changes on the soft tissues greatly alter the face. Even after-death masks do not retain the same facial expression as the living person had.

The Russian forensic medicine expert, Gerasimov, produced around 150 plastic reconstructions of historical personalities. Some authors dispute the scientific basis of his work. His portraits are rather freely artistic portrayals. In the cases of renowned personalities (Bach, Beethoven, Dante, Haydn, Kant, Leibnitz, Raphael, Richelieu, Schiller, Shakespeare), Welcker and others, on the basis of the skulls, produced numerous comparisons with extant posthumous masks, portraits or busts. They concluded that the profile of the death mask was not the same as that of the living man. The difference lies in the chin – usually the lower jaw is slightly open (Gerasimov 1967; Berg 1981; Prag, Neave 1989).

Gerasimov reconstructed the face in three stages: first, by anthropological research on the skull, then by graphical reconstruction, and finally by creating the sculpture. In his work he was aiming mainly to produce an approximate, not wholly complete likeness. This method is extremely demanding. The anatomist His, together with the painter Seffner (1895), composed an image of the soft facial parts for Bach's skull.

Likewise, the sketched reconstruction of the face is, in regard to likeness, scientifically questionable (Pusher, Szilvássy 1996). It requires not only a great deal of knowledge, but also

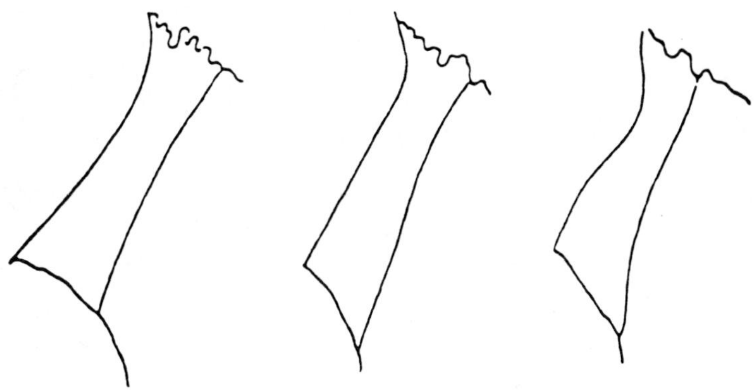

Picture 20: *The shape of the nasal bone determines whether the nose will be concave, straight or convex (Saller 1959)*

a fine sensibility in the fingers for modelling. The parallel projections, which are used for such reconstruction, were in our case unfeasible, since there were no extant original portraits. The preserved oil painting of Hermann II of Celje from the Croatian monastery of Lepoglava is, in the opinion of art historians (acad. E. Cevc), most likely produced after the stereotype model which was used for portraiture of the nobility in the 14th c. We also know that, among the extant skulls which supposedly belonged to the Counts of Celje, there is assuredly not the skull of Hermann II of Celje, since he was buried in the monastery church in Pleterje. Following the rearrangement of graves at the Pleterje village cemetery at the beginning of the 20th c. his grave remained unmarked and his posthumous remains were lost.

The present-day possibilities for facial reconstruction also include photographic reconstruction, which is based on comparing a photograph from a document and having the image superimposed on the skull, or by super-projecting a radiograph of the skull onto the photograph. Diverse authors have also referred to comparable research studies and to the associated problem areas (Eickstedt 1925; Stadtmuller 1961; Helwin 1971), computer reconstruction (Yoshino et al. 1997) and hologramic reconstruction (Harradine et al. 1980).

None of the above-mentioned theoretical methods of reconstruction was realizable in the present study.

Skulls of the Counts of Celje, preserved at the Church of Mary's Assumption in Celje. In 1956 they were transferred to the Celje Regional Museum

4.
RESULTS

4.1.
Historiographic Results

4.1.1.
Genealogical and Biographical Data for Those Presumably Buried in the Family Vault

The long passage of time has forever eradicated numerous traces. Yet the Counts of Celje are the only powerful representatives of the nobility who were conducting the economy on the territory of present-day Slovenia. This is why they are important for the Slovene people, and why their skulls have been protected. Many historians and others have been interested in them, and have recorded their life paths. Since it was the intention of this research to answer the question of who had been buried in the family vault, not only were all the secondary sources consulted, but also all the extant primary sources, the records of Celeian documents from the Staatsarchiv, Hofarchiv and Hausarchiv in Vienna (Otorepec). These were returned to Slovenia after the country gained its independence (Kos 1996). Further information was also sought in the Slovenian Archives, the Historical Archives in Celje, and also gathered from the church documents in Celje and from the monastery of Pleterje.

The starting-point of this study is the assumption that the male descendants lived on the home estate, while the females married elsewhere. Therefore it would be expected that the family tomb would contain the male representatives of the Counts of Celje, their wives and children.

The Counts of Celje – with the exception of Hermann II of Celje – had been buried between the years 1354 and 1456, i.e. for over a century, in the Minorite church of Mary's assumption (or St Mary's church) in Celje. The latter was built in 1242 by the Heunburg gentry (Marolt 1930). It contained three tombs: under the grand altar for the counts and princes of Celje; to the rear left for the Minorites, successors of St Francis of Assisi and by the entrance on the right for the Soteška gentry. St Mary's church burnt down twice, first in 1687, then again in 1811. Following the renovations after the second fire, the tomb of the Celeians was unprofessionally opened; however, to the credit of Janez Anton Zupančič, a teacher in a school of classical studies, the skulls were preserved behind the main altar. Altogether 18 skulls were preserved, but all the long bones were lost, as well as all the pelvic bones, the mandible – apart from half of one – and the teeth, apart from four. The superscription above the skulls read as follows:

Cellejensium Comitum ac Principum armis olim potentium omnibus fortunis abundantium pavcaerelquiae hic acquiescunt. Ludovicus imperator anno 1341 Fridericum Liberum a Sanegg primum Cellejensem Comitem creavit, cujus stirps anno 1456 interfecto Ulrico Comite interiit. (Herein lie the few mortal remains of the Counts and Princes of Celje, formerly wealthy and mighty in arms. Emperor Ludwig in 1341 named Friedrich I of Sannegg as the first Count of Celje. His lineage died out in 1456, when Ulrich was killed.)

On the skull, which is distinct from the others in that the left half of the face is missing, most probably the earlier mentioned teacher had affixed an inscription bearing the words:

Ultimus ex illustri familia Comitum de Cillie occisus per Ladislavum Hunyadi Belgradii 26. Martii 1456. (The last of the illustrious family of the Counts of Celje, whose murder was ordered by Ladislaus (László) Hunyadi in Belgrade on 26 March 1456.)

Around the year 1695 somebody erected a monument to the deceased counts and princes in St Mary's church, and attached it to the wall by the main altar on the women's side. The inscription read:

MONUMENTUM ILLUSTRISSIMORUM D.D., COMITUM CILLIENSIUM FUNDATORUM HUIUS BONUS MARIANAE, QUORUM CORPORA HIC SEPULTA SUNT, NOMINA AUTEM EORUM VIVUNT IN GENERATIONEM ET GENERATIONEM. (Monument to the most illustrious Counts of Celje, founders of this Church of Mary, whose bodies herein are buried, yet whose names live on from generation to generation.)

Then the names of the counts and princes are listed:

*Anno Domini 1360 vitam cum morte comutavit comes **Fridericus** 20. Maji hic sepultus.* (In the year of the Lord 1360 on 20th May life was exchanged for death by Friedrich here interred.)

*Anno Domini 1366 e vita migravit comes **Udalricus** 27. Jul. hic sepultus.* (In the year of the Lord 1366 departed from life Ulrich here interred on 27th July.)

*Anno Domini 1372 extremum diem clausit comes **Joannes** 18, Maji hic sepultus.* (In the year of the Lord 1372 Count Hans concluded his ultimate day and was here interred on 18th May.)

*Anno Domini 1382 fata solvit comes **Hermanus** 20. Martii hic sepultus.* (In the year of the Lord 1382 Hermann took leave of life and was here interred on 20th March.)

*Anno Domini 1392 valedixit orbi comes **Wilhelmus** 22. Sept. hic sepultus.* (In the year of the Lord 1392 Count Wilhelm bade farewell to the world and was here interred on 22nd September.)

*Anno Domini 1432 vivere cessavit comes **Hermanus** sepultus in Pletriarch.* (In the year of the Lord 1432 Hermann ceased to live and was interred at Pleterje.)

*Anno Domini 1443 expiravit comes **Georgius** 1. Maji hic sepultus.* (In the year of the Lord 1443 Count Jurij (Georg) expired and was here interred on 1st May.)

*Anno Domini 1444 finem vivendi fecit comes **Hermanus** 30. Jul. hic sepultus.* (In the year of the Lord 1444 Hermann concluded his life and was here interred on 30th July.)

*Anno Domini 1452 decessit ex hoc mundo comes **Hermanus** 5. Octob. hic sepultus.* (In the year of the Lord 1452 Hermann departed from this world and was here interred on 5th October.)

*Anno Domini 1453 decimatus a morte princeps et comes **Fridericus** 19. Jul. hic sepultus.* (In the year of the Lord 1453 death slew the Prince and Friedrich who was here interred on 19th July.)

*Anno Domini 1456 pie obiit princeps et comes **Udalricus** 12. Nov. hic seputus.* (In the year of the Lord 1456 in piety died the Prince and Ulrich, here interred on 12th November.)

Thesaurizaverunt sibi Thesaurum indeficientem in coelo, vade Christiane et fac tu similiter. (They have stored for themselves indestructible treasures in heaven. Go Christian, do thou likewise.)

This monument, with its numerous inconsistencies, remained in the church until the fire of 1811. Where it disappeared to thereafter is not known.

In anticipating the age of the individuals buried, we were frequently assisted by the dates of marriage, which are usually preserved in the documents. From them, it was in certain cases possible to conclude about the approximate dates of birth; with the exception of **Ulrich II** of Celje, the birth-dates could not be precisely established from the extant documents. During the Middle Ages, people married very young. The reasons for early marriage were several: short life-span; frequent illnesses and deaths caused by contagious diseases; high mortality rate of mothers in childbirth; and also, among representatives of the upper classes, fear of the demise of the family lineage and the desire to expand property and estates. Consequently, the firstborn children in a marriage were frequently conceived shortly after sexual maturity had been attained by the young mother (and often also the father). Given the above factors, it was presumed that the girls had married aged between 15 and 20, and the boys a few years older.

Friedrich I of Celje ordered the construction of the family vault; he died in 1359 (Krones 1972) or in 1360 (Grafenauer 1988). In 1341 (16. 4.) he was elevated to the title of Count of Celje. He certainly did not come to this flattering title at a very early age, although some credit for it is due also to his forebears. Since the available documents did not make it possible to establish his precise age, it was inferred. Ignac Orožen records that, in the year 1318, Friedrich was still of minor age. He lived for almost a further two decades following his ennoblement to the status of Count; it is therefore surmised that this elevation did not take place before he was aged thirty. Hence, one may conclude that Friedrich I would have reached an age of approximately 50 years or more. It is probable that he was the first male family member to be buried in the Celje vault.

In the majority of instances from the genealogical line of the Counts of Celje we do, in fact, know the marriage dates, since for weddings records of the occasion were kept, including dowry lists etc., and these documents are for the most part preserved – but there are none for Friedrich I. Most probably, in 1330, Friedrich I married **Didmund (Diemut) of Walsee**. She bore four children, two sons – Ulrich I of Celje and Hermann I of Celje – and the daughters Catherine and Anna. If we bear in mind the fact that in the late Middle Ages, marriages were contracted at a very early age and that children were already being born early in married life, then it is most likely that Didmund of Walsee lived at least till the age of 30. Possibly she was the first woman to be interred in the family vault. According to Orožen's Chronicle of Celje (Celjska kronika), Friedrich I may have been married twice: the first wife Ana, a Saxonian duchess, purportedly gave birth to both daughters, after which she presumably died, and therefore Friedrich remarried. The surmised first wife is not expected to be found among the collectively interred, since at the time of her death the family members were probably not yet being buried in the common vault.

Ulrich I of Celje died on 26 July 1368 (Krones 1972). The commentary to Krones's work (Krones 1972), maintains that he died when still young. He was probably born around the year 1330, or even before. Indications may be traced that he was already engaged in battle at Zadar in 1345, one year later in the Tyrol, in 1350 in Prussia and Lithuania; in 1354 he set off with the emperor Charles IV and his army for Rome, then one year later he was fighting with the Magyar-Croatian army against the Venetians, and in 1365 in Bulgaria. He was married to Adelaide of Ortenburg, the widowed countess of Ottingen (Krones 1972). They had a son, Wilhelm. Orožen writes, however, that Ulrich was married twice: first to Adelaide of Ottingen, with whom he is thought not to have had children; second, to Adelaide of Ortenburg. Here, most probably, an error is involved. Adelaide was soon again widowed (1368) and – according to extant sources – did not marry again.

Ulrich I of Celje was »an advocate« in the monastery of Žiče, which signified great political power, and therefore it would also be possible that he was buried in Žiče. In the monograph Carthusian Monasteries of Žiče and Jurklošter (»Kartuzije Žiče in Jurklošter«) (Mlinarič 1991), there is no record of this. The fact remains, however, that laypersons were not buried in monastery churches, with the possible exception of founders. This is why we are of the opinion that the couple was buried in the family vault in Celje.

How old was Ulrich I of Celje? Possibly 35 years or more. Among the extant documents, in the record from the day of 7 March 1385 (Otorepec, reg. I, no. 339), the name of Adelaide of Ortenburg in relation to brother-in-law Hermann was found.

After this year, Adelaide's name was no longer to be found in the documents. She died in 1391 and lived to see »mature years« (Krones 1972).

Hermann I of Celje probably died on 21 March 1386, on St Benedict's day (Krones 1972) or one year later (Orožen 1997, Grafenauer 1988). As for the others, for him as well the birth details are not given, so one can only surmise about his lifespan. Since his father (Friedrich I) lived relatively long – possibly around 50 years – and the son outlived the father by 25 years, one may suppose that Hermann himself also reached approximately the age of 50. It is to be expected that his remains were buried in the family vault.

Catherine Kotromanić was, according to the extant sources, the only wife of Hermann I of Celje. She bore him two sons, Hans and Hermann II of Celje. The precise years of her birth and death remain unknown, although according to the sources it may be concluded that she was born shortly after the year 1340 (she was married in 1360 or 1361) and died before 1408. Most probably, she did not die young, for this would in all likelihood have been recorded in the preserved sources, or else Hermann I of Celje might have remarried. In the register extracts (reg. e.) her name is also mentioned in an entry for 12 March 1387 (a deed with her son Hermann II of Celje and nephew Wilhelm), and the last mention of her name was found in the documents of 1401 (Otorepec, reg. e. no. 407). What age did Catherine Kotromanić live to see? Our estimate is over 60 years. One would therefore expect to find also her posthumous remains in the family vault.

Hans was the son of Hermann I of Celje; he died, still young, on 29 April 1372, leaving no heirs (Krones 1972). It is known, however, that in 1369 he was married to the countess Margarethe of Pfannberg. Three years after the marriage he died of unknown causes. How old was he? Given that his parents were married between the years 1360 and 1361, that he was their first child, and that he died in 1372, one may conclude that he must have been aged at most 11–12 years (Krones 1972). They must therefore have married him off as a child, aged 9 years at most (marriage in 1369). His remains are expected to be found in the family vault. Following Hans' early death, his widow was remarried to Count Montford; hence, she is not expected to be found in the collective vault.

Wilhelm was born in 1362. In 1392 he returned from a campaign against the Turks; in Vienna he fell ill, and died there on 19 September of the same year (Krones 1972). Although he died on foreign soil, one may believe that, since the tradition of burying the Celje nobility was already so well established 32 years after the vault had been constructed, he too would have been brought back from Vienna to be buried at home. Yet we are not aware of any written documents which might confirm or refute this assumption, and most probably none exist. According to a source from 14 October 1373 (Otorepec, Reg. II., no. 18), Wilhelm had been betrothed to Elisabeth, daughter of Meinhard from Carinthia. Clearly, however, a better opportunity arose, and in 1380 (Grafenauer 1988) Wilhelm was nobly married to Ana, daughter of the Polish King Kazimir Veliki Piast (Reg. II., no. 29 gives the date as 1382). He was married at the age of 18, and had a daughter, Anna. The widow Anna remarried and became Princess Teck. She died in 1425, and is buried in the church at Mindelheim (Orožen 1997).

Hermann II of Celje is indisputably buried in the monastery church of Pleterje (Mlinarič 1982). He was born around 1360, and while still a minor was married in 1372 to Anna of Schaunberg, daughter of Henrik of Schaunberg. In 1435, he died in Bratislava, and his body was taken to Pleterje. The birth year of his wife is not known, but she is presumed to have died before 1396 (Orožen 1997). Her age at death is also not known, however, given the fact that she was married in 1372 and gave birth to six children – the last being Barbara of Celje (between 1381 and 1385) – one may conjecture that she must have lived at least 40 years. Although no documentary record is available regarding her burial in the Celje family vault, it may be presumed that she died before the construction of the Pleterje chapel. Furthermore, in the monastery chapel the only laypersons who could be buried were exclusively the founders. It is not precisely known when Hermann II of Celje decided that he wished to be buried in Pleterje. He began building the Carthusian monastery in 1403.

Hermann III of Celje, son of Hermann II of Celje, is presumed to have been born between the years 1380 and 1385, which would be consistent with the fourth birth given by his mother, married in 1372. He was twice married: first in 1401 (Grafenauer 1988), and then (according to the same author) in 1404. An original document testifies that the second marriage was in 1424 (Otorepec, Reg. II., no. 71). This document, dated 9 February 1424, is the letter of betrothal between Hermann III of Celje and Beatrix of Bavaria (Ernest Count of Bavaria yields the hand of his daughter to the fiancé for 12,000 gulden.). The same year is also cited by Orožen. Hermann III of Celje had no male offspring. He died as a result of falling from a horse at the castle of Kamen (Stein) near Begunje in 1426 (Grafenauer 1988) or 1427 (Krones 1972). One may surmise, therefore, that at the time of his death he was aged between 35 and 45 years. His body was taken to Celje and interred in the family vault. One would expect to find there also the posthumous remains of his first wife – **Elisabeth of Abensberg**. It is not known precisely when she died. If one conjectures that Hermann III of Celje was not a widower for long, then his first wife would have lived at most to the age of 35–40 years. His second wife bore a daughter, Margarethe; during their marriage of two years they had no male offspring. Beatrix, as a rich widow, would probably have soon remarried, therefore one would not expect to find her in the family vault.

Ludwig of Celje was, likewise, the son of Hermann II of Celje. His birth-date is not known. Since Friedrich of Ortenburg did not have descendants, he came to an agreement with Hermann II of Celje that Ludwig be adopted. When this actually happened is not known, however it is known that he died of undetermined causes while still young, in 1417 (Grafenauer 1988) or, according to other sources, between 1414 and 1420 (Krones 1972). Since it is probable that he was born after Hermann III of Celje – and it is known that he died young – one may conclude that he did not even reach the age of 30. Sources relating to the betrothal and marriage have not been found. Nor is there any record among the extant sources of the place of his burial. It is highly probable that, after his death, he was removed either to the Celje vault or to the Ortenburg family tomb in Spittal (for example, Friedrich of Ortenburg, after his death, was also buried in the Ortenburg tomb in Spittal, and not in Radovljica, since that tomb was too small).

Hermann, illegitimate son of Hermann II of Celje, became bishop of Freising in 1412, and in 1421 bishop in Tridentum (Trent). In the same year he died. Since he was a legitimized illegitimate child, and since he was ordained for the spiritual life, he was probably not buried in the family vault. Or was he? There is, in fact, a plaque preserved in the church of St Daniel in Celje, which is just a few hundred metres distant from St Mary's church with the family vault; this funereal plaque, however, is built into the wall by the church doors, thus indicating that it had been brought from somewhere else. In the Church of St Daniel, however, there is no grave.

Friedrich II of Celje, son of the wedded couple Hermann II of Celje and Anna of Schaunberg, born in the 1380s, died in 1454. He lived till around the age of 80. He is interred in the family vault.

Elisabeth of Frangepan, murdered in 1422 or 1423 in Krapina, is also believed to have been buried in the family vault. The Chronicle records that she was brought dead from Krapina to »the monastery in Celje« (Krones 1972). This is a reference to the minorite monastery in Celje (Gubo 1909). She had been betrothed to Friedrich II of Celje on 30 September 1388 (Otorepec, Reg. II, no. 35), »however, on account of the tender age of the affianced, the marriage ceremony was conducted much later«, though before the year 1406, when Ulrich II of Celje was born. The marriage lasted 15 to 20 years. If one estimates that, at the time of the marriage, Elisabeth would have been aged between 15 and 20, then she would have reached an age of between 30 and 40 years.

Friedrich III, son of Friedrich II of Celje, died as a child, but it is not known how old he was. Most probably he was buried in the family vault; it is, however, possible that because of the extremely delicate child bones this skull was not preserved.

Hans was the **illegitimate son of Friedrich II** of Celje, hence the half-brother of Ulrich II of Celje. He was legitimized on 15 November 1447. Nothing is known of his biographical details. Since he was legitimized and also entered into all genealogical trees, one might expect to find him in the vault.

Ulrich II of Celje was born in 1406, and died on 9 November 1456. He was buried in the Minorite church in Celje. His children – Hermann IV of Celje, Georg and Elisabeth – all died as minors.

Hermann IV of Celje, first son of Ulrich II of Celje, according to the genealogical tree of C. Thomas (Thomas 1991), was born between the years 1432/33 and, according to the same author and to Krones, died on 30 July 1444 (Krones 1972), aged 11. However, in other sources (Grafenauer 1988) we find the date of his death given as 1452. Vlado Habjan writes that he was born in 1439. It is anticipated that his posthumous remains will be found in the shared vault.

Georg, second son of Ulrich II of Celje and Catherine Branković, was the first of their children to die; according to some sources, in 1443 (Grafenauer 1988), and to others, on 1 May 1424 (Krones 1972); the latter date is surely erroneous, since Ulrich II of Celje and Catherine Branković were married only in 1433. The Chronicle maintains that Georg died while still in the cradle. In this case, one would not expect his posthumous remains to be preserved.

Elisabeth was the third child of Ulrich II of Celje. Most of the sources do not mention when she was born. V. Habjan gives the year 1441, while C. Thomas in her genealogical record gives the birth year as 1443/44, and the year of death as 1455. Amongst other details in the same record, she writes that on 1 August 1455 Elisabeth gave birth to a son, Ladislaus (László) Hunyadi. It is difficult to believe in this birth year, since the 11–12-year-old girl was truly too young for motherhood! She expired »in her young days« ... »still too immature for marriage with Mátyás, son of the Magyar leader János Hunyadi« (Krones 1972). Since 1451 she had been betrothed to him (Grafenauer 1988). Ulrich II of Celje had in fact forced Hunyadi János to engage his son to Elisabeth, while for this purpose he was annulling her engagement to the Gorizian Count Johann (Krones 1972). The document of 11 December 1443 is a marriage contract between the minor-aged Elisabeth and the Gorizian Count Johann (Otorepec, Reg. II., no. 120). In the opinion of certain historians, she had been married just three months and would, of course, have been living with the Hunyadi family in Varaždin. It is conjectured that Elisabeth did not even reach the age of 15, and hence that she would be buried in the Hunyadi family tomb in Varaždin.

Ulrich's widow, Catherine Branković, lived in the castle in Krško for several years after her husband's death, battling for the Celje inheritance; she then travelled to Dubrovnik, then to Corfu, back to Belgrade, and finally settled in Ježevo in Macedonia, where she lived with her sister Mara, a widower from the campaigns of the Turkish Sultan Murat V. She died at a very advanced age and was buried at Konća, west of Strumica (Krones 1972).

4.1.2.
Historically expected interments in the family vault

4.1.2.1. The interment chronology of male members of the family of the Counts of Celje

1. 1359 or 1360 – Friedrich I
2. 1368 – Ulrich I of Celje
3. 1372 – Hans
4. 1385 – Hermann I of Celje
5. 1392 – Wilhelm
6. 1417 – Ludwig (adoptive son of Ortenburg)
7. 1421 – ? Hermann, Bishop of Freising, legalised son of Hermann II of Celje
8. 1426 or 1428 Hermann III of Celje
9. after 1447 – Hans, legitimized son of Friedrich II of Celje
10. 1444 or 1452 Hermann IV of Celje
11. 1454 Friedrich II of Celje
12. 1456 Ulrich II of Celje

4.1.2.2. The interment chronology of female members of the family of the Counts of Celje

1. 1353 or 1354 – Didmund of Walsee
2. 1391 – Adelaide of Ortenburg
3. before 1408 – Catherine Kotromanić
4. before 1396 – Anna of Schaunberg
5. 1422 or 1423 – Elisabeth of Frangepan
6. before 1424 – Elisabeth of Abensberg

Since there are mostly no recorded years of death for the women, the chronological order of their interments can be only presumed.

4.1.2.3. Common interment chronology of all members of the family

1. (F) 1353 or 1354 Didmund (Diemut) of Walsee
2. (M) 1359 or 1360 Friedrich I
3. (M) 1368 Ulrich I of Celje
4. (M) 1372 Hans
5. (M) 1385 Hermann I of Celje
6. (F) 1391 Adelaide of Ortenburg
7. (M) 1392 Wilhelm
8. (F) before 1408 Catherine Kotromanić
9. (F) before 1396 Anna of Schaunberg
10. (M) around 1410 Friedrich III (baby)
11. (M) 1417 Ludwig (adoptive son of Ortenburg)
12. (M) 1421 Hermann? (Bishop)
13. (F) 1422 or 1423 Elisabeth of Frangepan
14. (F) before 1424 Elisabeth of Abensberg
15. (M) 1426 or 1428 Hermann III of Celje
16. (M) 1443 Georg (baby)
17. (M) after 1447 Hans (legalized son of Friedrich II of Celje)
18. (M) 1444 or 1452 Hermann IV of Celje
19. (M) 1454 Friedrich II of Celje
20. (M) 1456 Ulrich II of Celje

Legend: (M) = *male,* (F) = *female*

4.1.3.
Anticipated age of the interred in the family vault

1. Friedrich I, 50 years or more
2. Didmund (Diemut) of Walsee, supposedly 30 years or more
3. Ulrich I of Celje, 35 years
4. Adelaide of Ortenburg, supposedly at least 20 years or more
5. Wilhelm, supposedly 30–40 years
6. Hermann I of Celje, about 50 years
7. Catherine Kotromanić, supposedly 30 years or more
8. Hans, supposedly not more than 11–12 years
9. Anna of Schaunberg, supposedly 40 years or more
10. Friedrich II of Celje, about 80 years
11. Elisabeth of Frangepan, 30–40 years
12. Hermann III of Celje, 35–45 years
13. Elisabeth of Abensberg, supposedly 30–40 years
14. Ludwig, supposedly not more than 30 years?
15. Hermann?, legitimized son of Hermann II of Celje, 40 years at most
16. Hans, legitimized son of Friedrich II of Celje?
17. Ulrich II of Celje, 50 years
18. Hermann IV of Celje, supposedly 11 years

4.1.4.
Members of the family, interred elsewhere

- The first members of the Family of Sannegg are interred in the monastery of Gornji Grad,
- Hermann II of Celje at Pleterje,
- Barbara of Celje at Melnikov in Bohemia,
- Veronika of Desenice first at Braslovče, later at the monastery Jurklošter,
- Catherine Branković at Konća in Macedonia.

Other female descendants who married elsewhere, were also buried there.

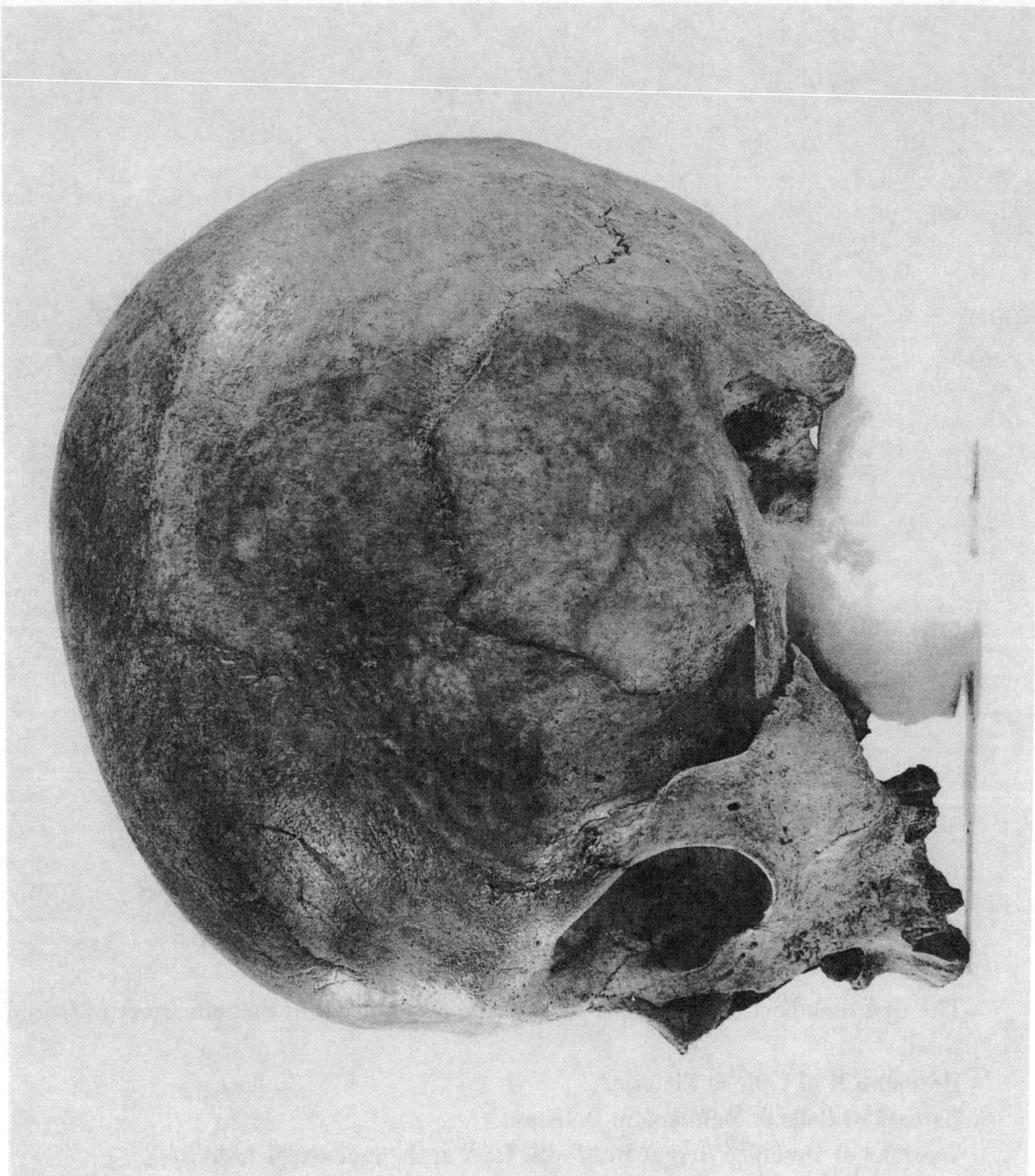

Picture 21: *Skull 1, side view. Visible is the protrusion of the maxilla*

4.2.
Identification Results

4.2.1.
Gender, Age and Illnesses by Skulls 1–18

4.2.1.1.
Skull 1 (pict. 21)

4.2.1.1.1. Determining gender by skull

Anthropological characters for determining gender

superciliary arches	slightly developed (–1)
glabella	moderately expressed (0)
frontal tubers	moderately expressed (0)
inclination of the forehead	more inclined (+l)
supraorbital margin	expressed (0)
orbital shape	expressed rounded (0)
occipital relief	averagely expressed (0)
ext. occipital protuberance	unexpressed (–2)
mastoid processes	relatively large (+1)
cheekbones	moderately expressed (0)
maxillary palatine process	ellipsoid
maxillary alveolar process	between ellipsoid and semi-circular
thickness of calvaria	greater frontally than occipitally

Determining gender by anthropometric methods and indices

bizygomatic breadth	132 mm
great occipital foramen	(rounded), length 37 mm, breadth 31 mm
mastoids	both strong, breadth r. 45 mm, l. 30 mm
breadth of anterior nasal aperture	26 mm
cranial length	174 mm
cranial breadth	142 mm
cranial base	98 mm
Campers' angle	80° (protrusion of intercanine sector)
cranial index	81.61 (short skull)
orbital height	r. 36 mm, l. 35 mm
orbital breadth	r. 38.5 mm, l. 38.5 mm
orbital index	r. 93.5, l. 90.90 (female feature)

Determining gender by dental methods

Dental index after Pont:

Breadth between first premolars	35 mm. P = 43.75
Breadth between first molars	45 mm. M = 75

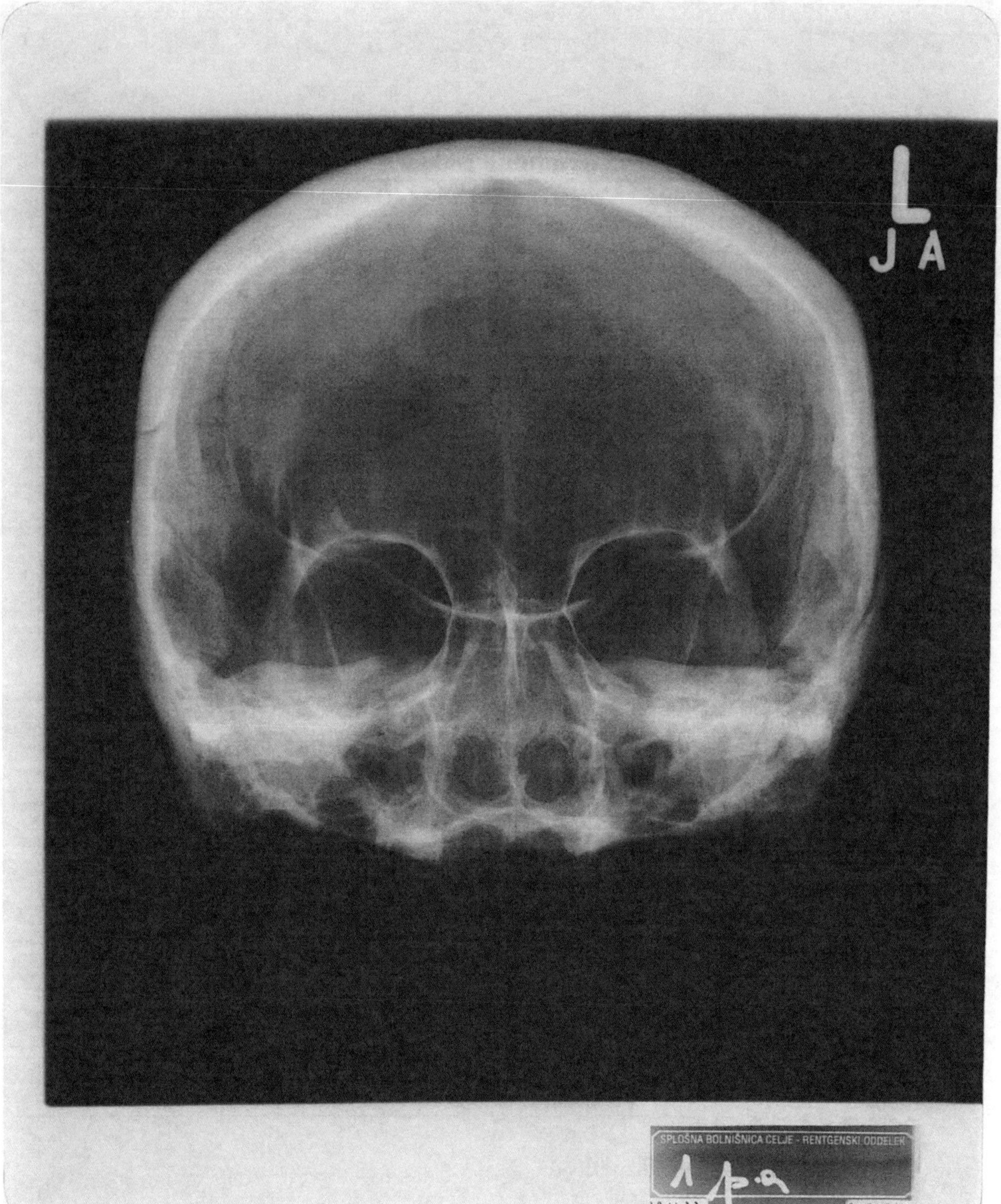

Picture 22: *Occipito-frontal radiograph of skull 1*

4.2.1.1.2. Determining age by skull

Determining age by endo- and ectocranial obliteration of the sutures of the skull

Among the cranial sutures, the sagittal suture is closed, the coronal suture in the temporal part is still not completely obliterated, the lambdoidal is obliterated in pars lambdica and pars media, though it still remains open in the asterical part. The medial palatine suture and temporal zygomatic sutures, both left and right, are well visualized and not obliterated.

Determining age by forensic-dental methods

The maxilla does not retain any preserved teeth. The medial palatine suture is closed, while the right palatine torus is pronounced. The incisive foramen is clearly expressed. The left tuber is broken off, the left dental alveoli are empty from 1 to 6.

On the right, the dental alveoli are more pronounced, and at the place of the second incisive, the canine and the first premolar, the maxillary processus alveolaris is broken off. The alveolus of the first canine is preserved. The alveolus of the second premolar is atrophic, therefore the tooth had fallen out during lifetime, while the alveoli of the first and second molars remain evident.

On the left is the dental alveolus of the first canine, but the alveolus of the second canine is broken off; the alveolus of the canine is also missing.

The dental alveoli of the first and second molars are preserved. The alveolus of the first molar is most unpronounced; possibly this tooth had already been lost during the lifetime. At the place of the second and third molars there are no dental alveoli; in the bone there is a circular defect – a fistula – which may be an indication of a chronic process (a granuloma or a cyst), possibly the second molar was roughly extracted or else abruption of the bone occurred.

Maxillary alveolar process from right towards left:

	M3	M2	M1	PM2	PM1	C	I2	I1	I1	I2	C	PM1	PM2	M1	M2	M3
Maxilla:	–	+	+	+a	+b	+b	+b	–	+	+b	+b	+	+	+a	+d	+d

Legend: + = *alveolus present,* – = *no alveolus, a = atrophic, b = broken off, d= bone defect*

Determining age by skull using X-ray methods (pict. 22)

On the side projection, there are visible remains of the lambdoidal suture, which in the asterical part is still not obliterated. The sella turcica is very deep, and the grooves are prominently visible on account of ramification of the meningeal artery. Asymmetry of the oval foramen is also visible; the right is considerably greater than the left. The Pacchionian bodies are within normal limits. The other radiographs are without special features.

No. 1: ♀, over age 50

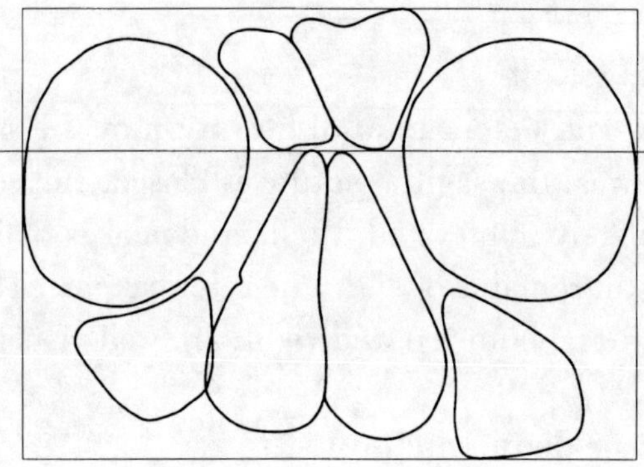

Picture 23: *Outline sketch of the facial cavities of skull 1 from the occipito-frontal radiograph*

4.2.1.1.3. X-ray determination of the facial cavities (pict. 23)

On the occipito-frontal radiograph one may see that:

- the two frontal sinuses are smaller, non-symmetric, and that the left is larger than the right. They have fanlike shapes;
- the external margins of the ocular cavities (aditus orbitae, hereafter the orbita) are of a more rounded shape, assessed after Nemeskéri as (0). Their surface is uniform, left and right;
- the maxillary sinuses are relatively small and non-symmetric, the left being larger than the right. In shape they are closest to triangular;
- the nasal bone is short, broad and relatively high. The nasal cavity is more pear-shaped in form.

Planimetric values of the facial cavities

Skull		**Frontal sinuses**		**Orbits**		**Maxillary sinuses**		**Nasal cavity**
		Right	**Left**	**Right**	**Left**	**Right**	**Left**	
No. 1	S	2.67	4.72	17.10	17.10	2.76	3.74	14.07
	H	2.32	2.85	4.81	4.81	2.14	2.85	5.52
	W	1.34	2.23	4.19	4.30	1.52	1.87	3.56
	T	0.54	0.72	0.81	0.81	0.63	0.54	0.54
	Σ	7.39						
	Sh	0.72	0.63	0.81	0.81	0.72	0.72	0.63

Legend: *S = surface, H = maximal height, W = maximum width, T = W/H, Σ = total surface of frontal sinuses, Sh = shape of sinuses*

4.2.1.1.4. Paleopathological methods

The mandibular fossa is eroded on left, smooth on right. It is possible that arthropathy occurred due to the early loss of teeth. The bone is also porous on account of osteoporosis.

On the radiograph in side projection, below the pterygoid fossa is a structure, well demarcated, which might have been an inborn cyst or haemangioma; this suspicion was excluded by subsequent X-ray analysis.

4.2.1.1.5. Commentary

According to the appropriate skull measurements and cranial index the skull is short, relatively heavy, with a bone thickness of less than 5 mm. It is undamaged. In its features it is not a typical female skull; the mastoids are somewhat larger, the forehead is more inclined and the cheekbones are stronger. The other features are typically female: the glabella is not pronounced, there are almost no superciliary arches or frontal tubers, the occipital protuberance is unpronounced, and the attachments of the occipital muscles are slight. The orbits are more rounded, and the orbital indices are indicative of the female gender. The nasal bone is pronounced, short and wide, the nasal septum straight. Campers' angle indicates the protrusion of the maxilla and the somewhat foreward projecting teeth. The dental index after Pont attests to an ellipsoidal palate and alveolar process of the maxilla; together they indicate a narrowish face. The dental alveoli are poorly preserved due to post mortem damages; at the place of the second and third molar right there is a round-shaped defect, which is indicative of a chronic inflammatory process, granuloma or cyst, as a result of which greater fistulation occurred. The left jaw joint reveals stronger degenerative changes, while osteoporosis is also evident elsewhere on the skull. Given the details recorded, it is anticipated that this is the skull of an elderly, fairly robust woman.

Particular feature: the palate is asymmetric, narrowed on the side, reaching to the alveoli of the first molars, where it is broken off. The palatine torus remains pronounced. The left palatine tuber is broken off. The maxilla in the intercanine sector is forward protruding, at an angle of approximately 80 degrees.

Although some of the features are ambivalent with regard to gender, most of the features – including also the orbital indices – indicate this as being a skull of female gender.

THE SKULL BELONGED TO AN ADULT WOMAN, OVER 50 YEARS OF AGE

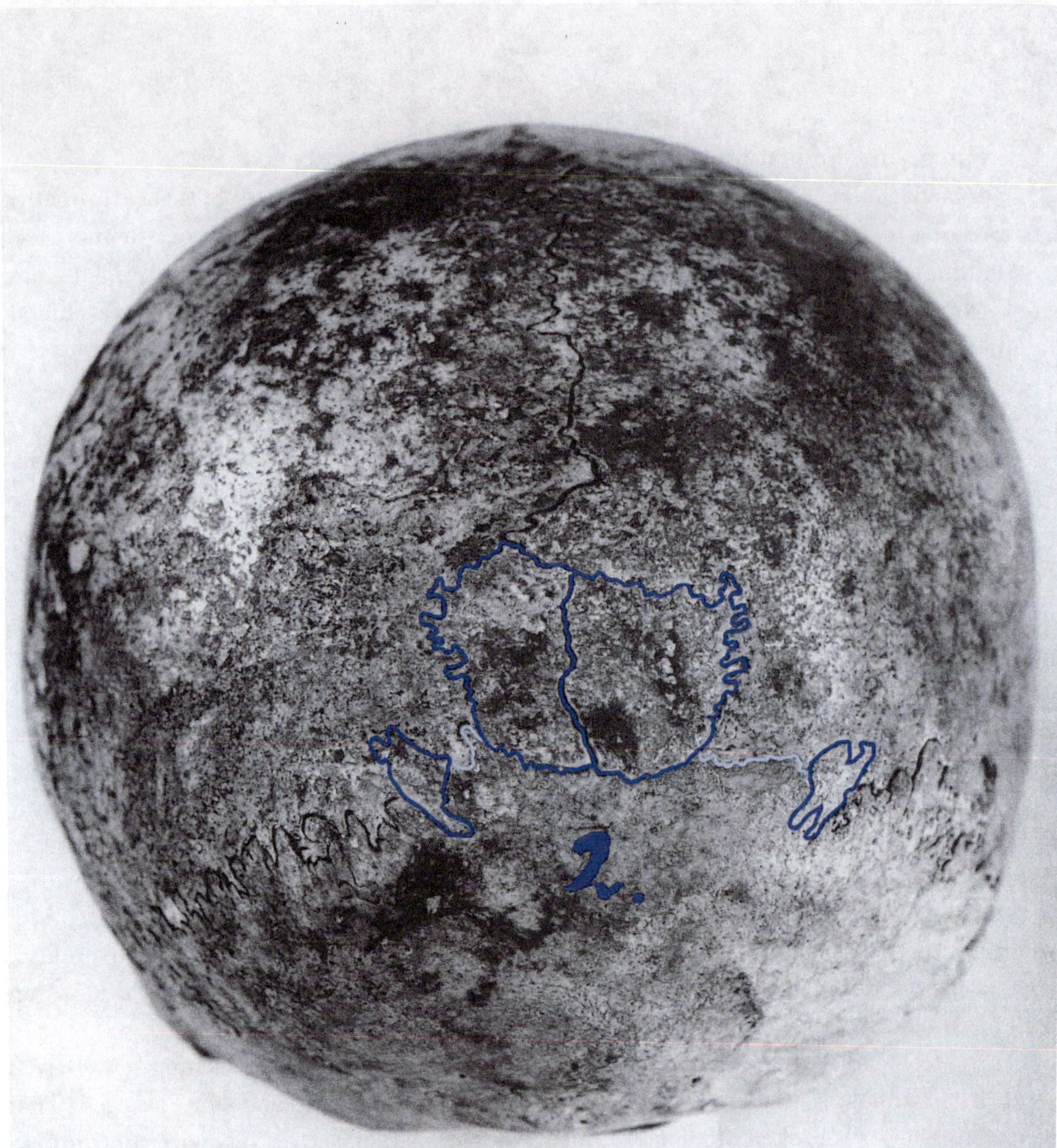

Picture 24: *Child's skull 2, from behind. Similarly as in this example, all the skulls are numbered at the cranial base. Four additional ossicles are visible: the larger lambdoidal and the smaller left and right in the lambdoidal suture*

4.2.1.2.
Skull 2 (pict. 24)

4.2.1.2.1. Determining gender by skull

Anthropological characters for determining gender

superciliary arches	not developed (–2)
glabella	unexpressed (–2)
frontal tubers	expressed (–1)
inclination of the forehead	almost vertical (–2)
supraorbital margin	sharp
orbital shape	rounded (–2)
occipital relief	unexpressed (–2)
ext. occipital protuberance	unexpressed (–2)
mastoid processes	small(–2)
cheekbones	both damaged
maxillary palatine process	ellipsoid
maxillary alveolar process	semicircular
thickness of calvaria	very thin
facial part of skull	small and narrow

Determining gender by anthropometric methods and indices

bizygomatic breadth	damaged
great occipital foramen	length 34 mm, breadth 28 mm
mastoids	both broken off
breadth of anterior nasal aperture	22 mm
cranial length	166 mm
cranial breadth	136 mm
cranial base	89 mm
Campers' angle	approx. 850
cranial index	81.92 (short skull)
orbital height	l. 33 mm, r. orbital surface of maxilla is broken off
orbital breadth	l. 35 mm, r. orbital surface of maxilla is broken off
orbital index	left – 87.71, r. undeterminable

Determining gender by dental methods

Dental index after Pont:

Breadth between first premolars	33 mm. P = 41.25
Breadth between first molars	45 mm. M = 75 regarding the medial palatine suture

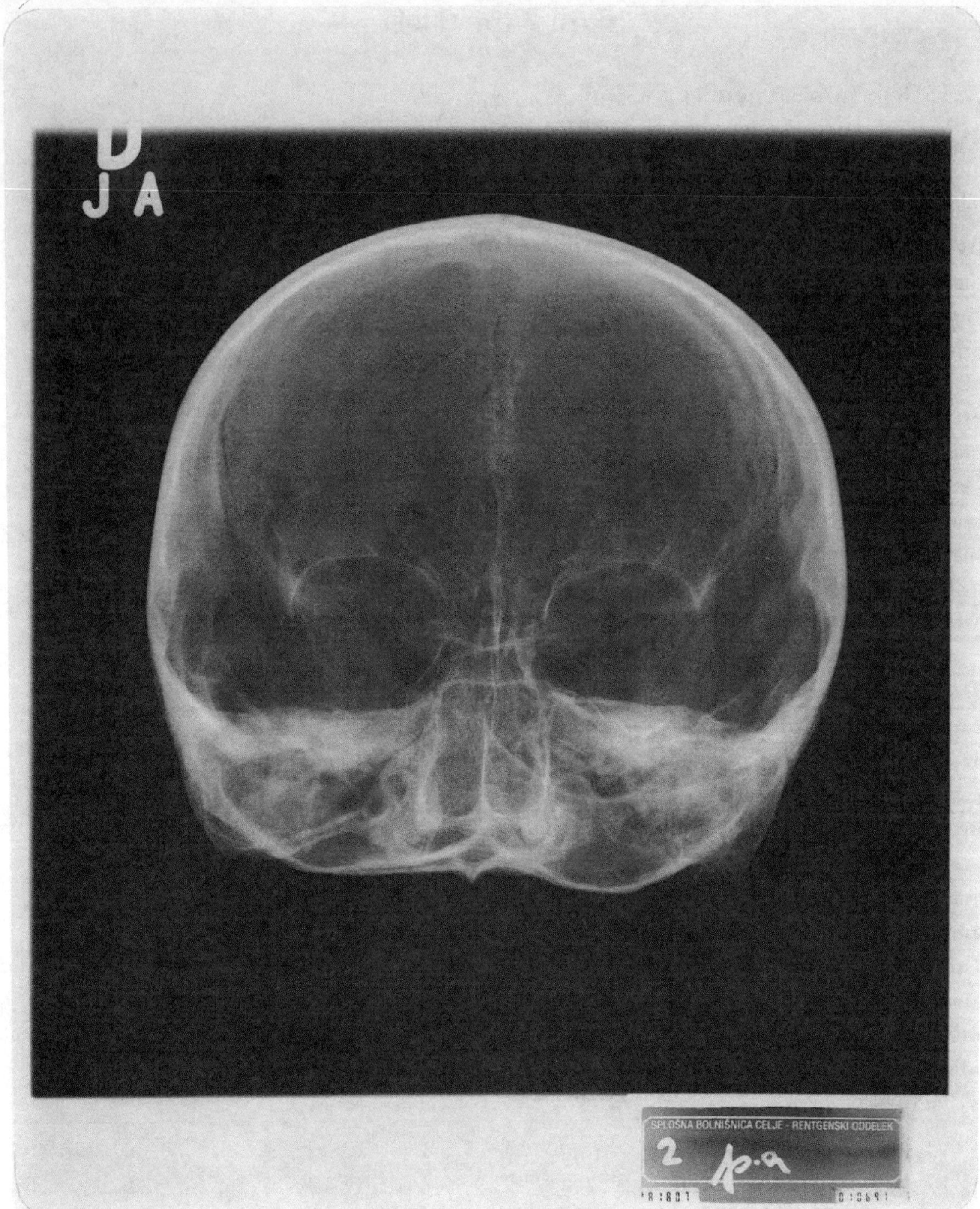

Picture 25: *Occipito-frontal radiograph of skull 2*

4.2.1.2.2. Determining age by skull

Determining age by endo- and ectocranial obliteration of the cranial sutures

The sphenooccipital synchondrosis is wide open. The sagittal, coronal and lambdoidal sutures are all open. The same holds for the medial palatine suture and for the left temporozygomatic suture. The state of the cranial sutures described indicates the skull of a child, aged around 10 years.

Determining age by forensic-dental methods

The left maxilla is preserved in its entirety; on the right the whole facial side is missing. The alveolar part left is well preserved, the right is damaged. The alveolar process of the right maxilla is broken off from the first premolar backwards. At the place of the first and the second premolara, left and right, both maxillary tubers are broken off. The maxilla has two teeth preserved. The incisive foramen is fairly pronounced. The palatine bone is pronounced behind the transverse palatine suture.

Maxillary alveolar process

On the right side the canine tooth is retained, and the empty spaces of the not yet grown first and second molars are also preserved.

On the left side, the alveoli of the first two incisors are preserved, the canine tooth is retained, the first premolar is still retained, the alveoli of the second and first premolars are also preserved, while the second molar still remains in the bone.

Maxillary alveolar process from right towards left:

	M3	M2	M1	PM2	PM1	C	I2	I1	I1	I2	C	PM1	PM2	M1	M2	M3
Maxilla:	b	b	b	b	+m(r)	+m	+m	+m	+m	+m	+t	+(r)	+	+	–	–

Legend: *t = tooth, b = broken off, m = milk-tooth, r = retained permanent tooth*

(right: I II III IV (4 retained), broken off left: I II III – tooth IV (4 retained) 5 6).
The dental condition described indicates a child, aged 9–11 years.

Determining age by skull using X-ray methods (pict. 25)

The radiographs reveal a gracile bone structure. The thickness of the vault of the skull (calvaria) is greater occipitally than frontally. Occipitally there are evident digital impressions, typical of a child's skull.

On both sides of the maxilla, there are evident traces of the beginnings of the first premolar. No other teeth are seen on the radiographs. The orthopan tomography provided technically unsuitable images.

The remaining radiographs of the skull showed no special features.

No. 2: ♂, 12 years

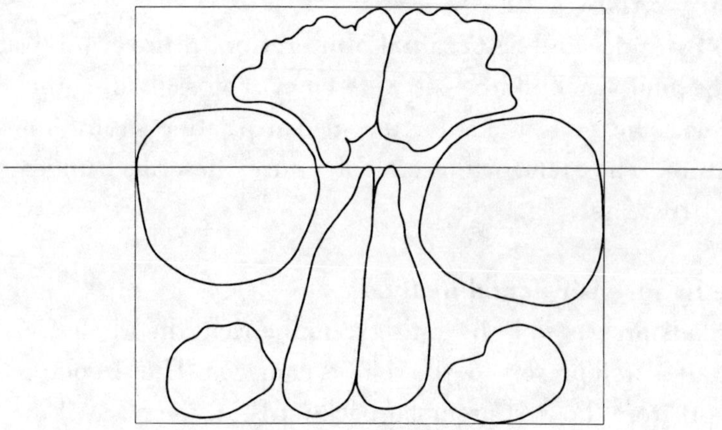

Picture 26: *Outline of the facial cavities of skull 2 from the occipito-frontal radiograph*

4.2.1.2.3. X-ray determination of the facial cavities (pict. 26)

On the occipito-frontal radiographs one may see that:

- the frontal sinuses are asymmetrical and fan shaped. Since this is a child's skull, and the two sinuses are still developing, they would most probably be much larger in an adult person;
- the two orbits are asymmetrical, the right is more rounded, the left more trapezoidal. After Nemeskéri they are scored as (–2);
- the maxillary sinuses are relatively symmetrical. They are leaf-like in shape;
- the nasal cavity is symmetrical. It is trapezoidal in shape.

Planimetric values of the facial cavities

Skull		Frontal sinuses		Orbits		Maxillary sinuses		Nasal cavity
		Right	Left	Right	Left	Right	Left	
No. 2	S	4.99	7.66	11.13	11.22	9.53	9.44	9.79
	H	3.48	3.83	3.74	3.83	4.19	3.83	4.01
	W	1.87	2.94	3.39	3.39	2.94	2.94	2.76
	T	0.45	0.72	0.81	0.81	0.63	0.72	0.63
	Σ	12.64						
	Sh	0.63	0.63	0.89	0.89	0.81	0.81	0.81

Legend: *S = surface, H = maximal height, W = maximum width, T = W/H, Σ = total surface of frontal sinuses, Sh = shape of sinuses*

4.2.1.2.4. Paleopathological methods

The occipital bone is thickened. This is well seen on the radiograph: the child most probably suffered from rickets. The mandibular fossa is without particular features on both sides.

There are no other particular features on skull 2.

4.2.1.2.5. Commentary

The skull is short, gracile, small and very light; the bone is only a few mm in thickness. The larger injuries on the left side of the face occurred after death. Missing features are the right cheekbone and left cheekbone arch, most of the nasal bone with the nasal septum and all of the facial side of the right maxilla. The orbits are trapezoidal, and the maxillary part of the right orbit is missing. There is no glabella, the superciliary arches are undeveloped, and the frontal tubers just indicated. The tops of both mastoids are damaged. The great occipital foramen is more oval shaped, the occipital protuberance is weakly pronounced, and the occipital condyles are considerably furrowed. The dental arch is semicircular, and within it the milk teeth are interchanged with the permanent teeth; they fell out after death. All four incisors and both of the first molars are already permanent teeth. The permanent canine and the second molars are still remnant in the bone. Considering the dental state, we conclude that the child must have been about 10 years old.

All the cranial sutures are still open, and in particular there is the central indicator for a juvenile skull - spheno-occipital synchondrosis. The ossicles in the lambdoidal suture are also worth mentioning here.

On the skull there is no evidence of any disease causes for the early death. Furthermore, on account of the poorly developed gender dimorphism, we still cannot with certainty determine the gender.

THE SKULL BELONGED TO A CHILD, AGED 12 YEARS

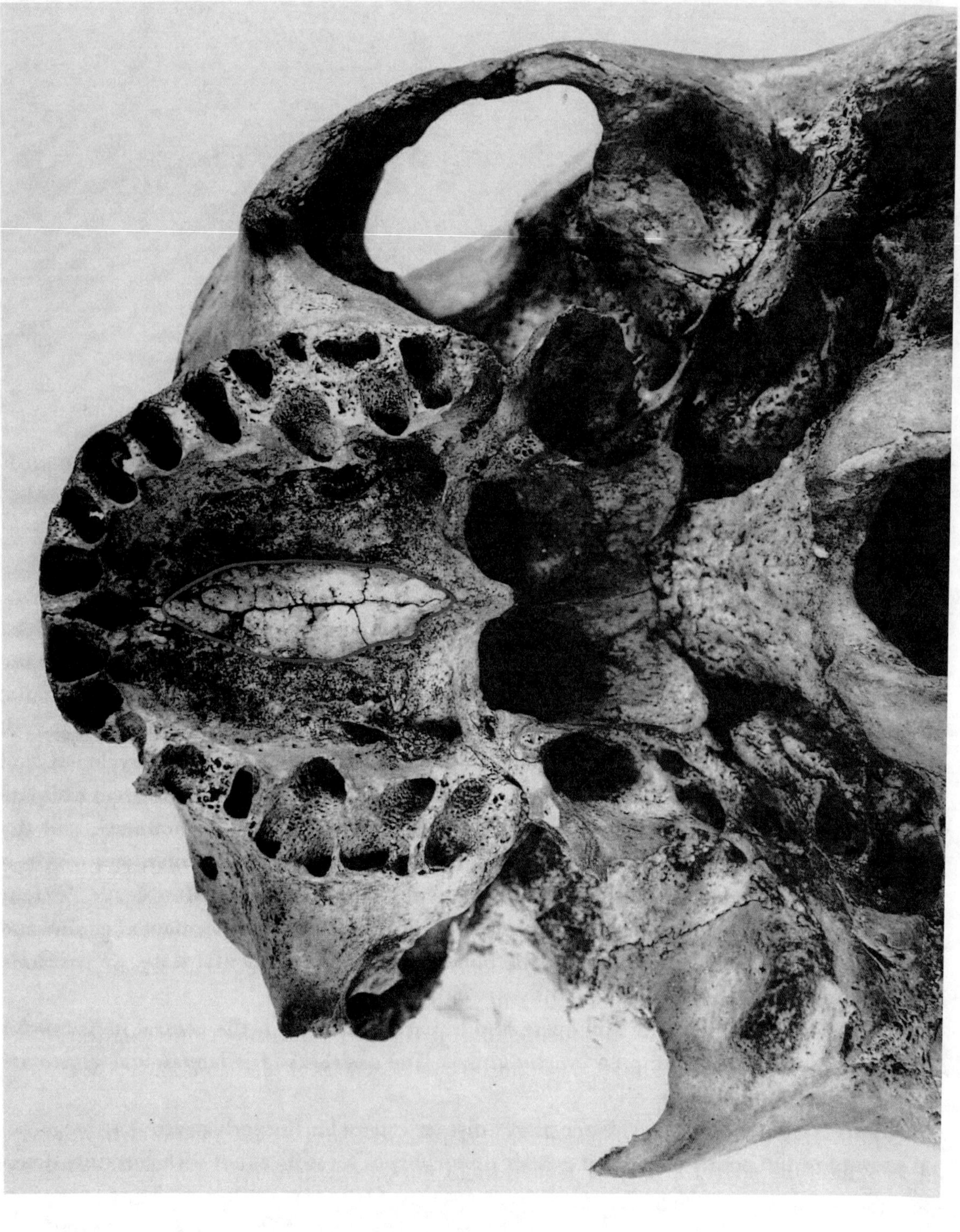

Picture 27: *Skull 3 from below. The dental arch is pronouncedly semicircular, all the tooth sockets are preserved; a particular feature is the highly pronounced palatine torus*

4.2.1.3.

Skull 3 (pict. 27)

4.2.1.3.1. Determining gender by skull

Anthropological characters for determining gender

superciliary arches	slightly developed (–1)
glabella	poorly expressed (–1)
frontal tubers	moderately expressed (0)
inclination of the forehead	almost vertical (–1)
supraorbital margin	indicated (0)
orbital shape	between rectangular and trapezoid (+1)
occipital relief	medium expressed (0)
ext. occipital protuberance	medium expressed (0)
mastoid processes	relatively large (+1)
cheekbones	r. broken off, l. moderately expressed
maxillary palatine process	ellipsoid
maxillary alveolar process	semicircular
thickness of calvaria	even

Determining gender by anthropometric methods and indices

bizygomatic breadth	? (right cheekbone broken off)
great occipital foramen	length 32 mm, breadth 27 mm
mastoids	large, breadth r. 40 mm, l. 30 mm
breadth of anterior nasal aperture	24 mm
cranial length	170 mm
cranial breadth	142 mm
cranial base	95 mm
cranial index	83.52 (short skull)
orbital height	r. not preserved, l. 34 mm
orbital breadth	r. not preserved, l. 38.5 mm
orbital index	left 88.31 (female features), r. undeterminable

Determining gender by dental methods

Dental index after Pont:

Breadth between first premolars	40 mm – 20 mm. P = 50
Breadth between first molars	49 mm. M = 81.66

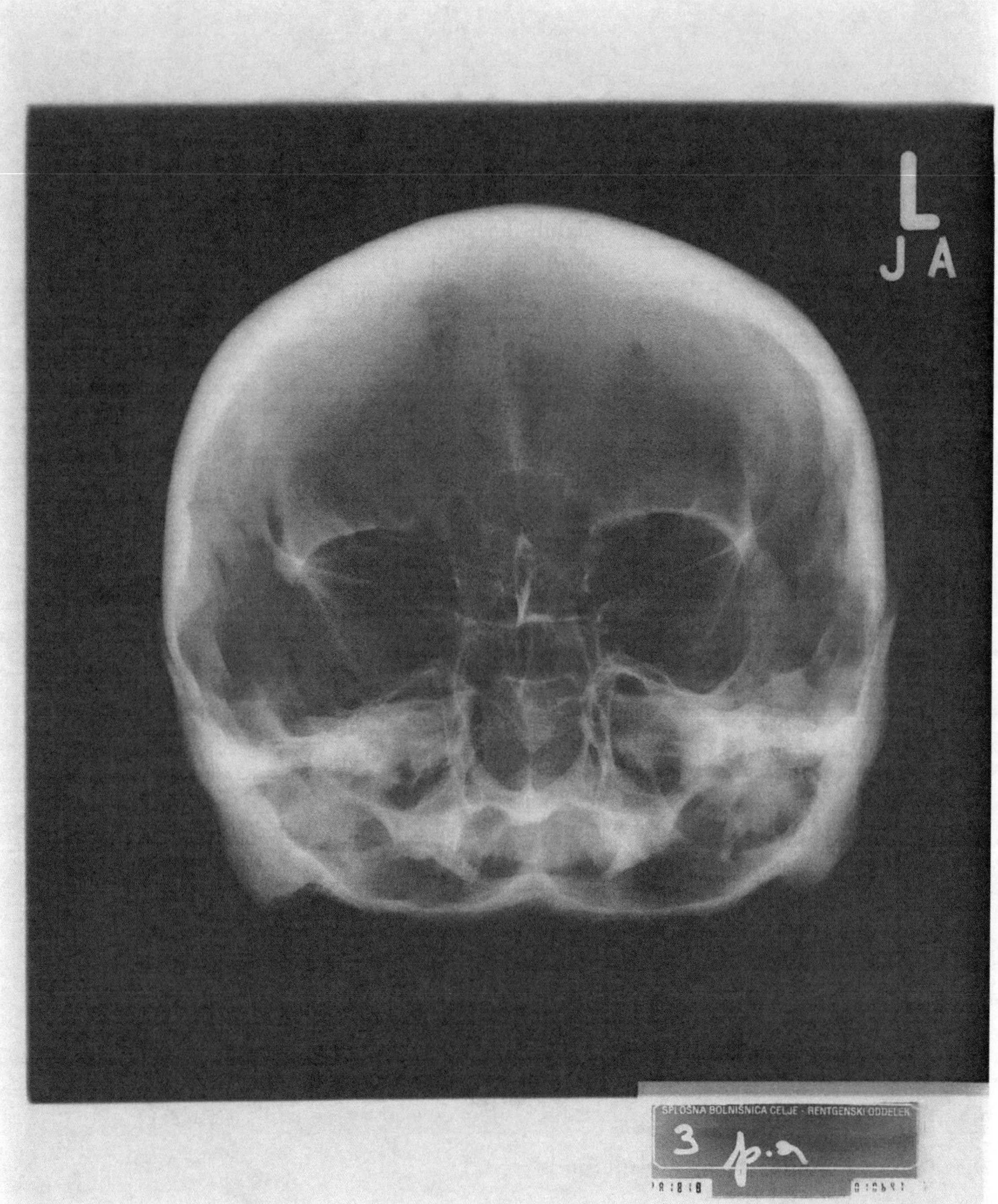

Picture 28: *Occipito-frontal radiograph of skull 3*

4.2.1.3.2. Determining age by skull

Determining age by endo- and ectocranial obliteration of the cranial sutures

The cranial sutures indicate that the process of obliteration has still not been concluded. The sagittal suture is still open in the pars postica, the coronal in the temporal part, and the lambdoid suture is still fully open. Likewise, the spheno-occipital suture is still not closed. The coronal suture cannot be seen; this may be attributable to hyperostotic superposition.

The features described indicate a person aged 30 years or more.

Determining age by forensic-dental methods

The maxilla retains no preserved teeth, however all eight dental alveoli are preserved left and right. All the hard palate is preserved. The incisive foramen is strongly pronounced. The median palatine suture is not obliterated; paramedially on both sides the palatine torus is strongly pronounced.

Maxillary alveolar process from right towards left:

	M3	M2	M1	PM2	PM1	C	I2	I1	I1	I2	C	PM1	PM2	M1	M2	M3
Maxilla:	+	+	+	+	+	+	+	+	+	+	+	+	+	+	+	+

Legend: *+ = alveolus present*

Determining age by skull using X-ray methods (pict. 29)

On the occipito-frontal radiograph we can see the calvary, which is frontally thickened in the sense of frontal internal hyperostosis (hyperostosis frontalis int.). The nasal bone is without particular features. The sagittal suture in the lower part is still not completely obliterated, the coronal suture is covered by the superposition of the frontal internal hyperostosis, and the lambdoidal is still fully open. The Pacchionian bodies are far more pronounced parietally to the right than to the left.

On the occipito-frontal radiograph right, paramedially and at the back one can see a circular formation 0.5 cm in size, which does not appear to be osteolysis; a macroscopic examination of the skull reveals a smaller artificial aperture at the same place.

No. 3: ♀, 30–40 years

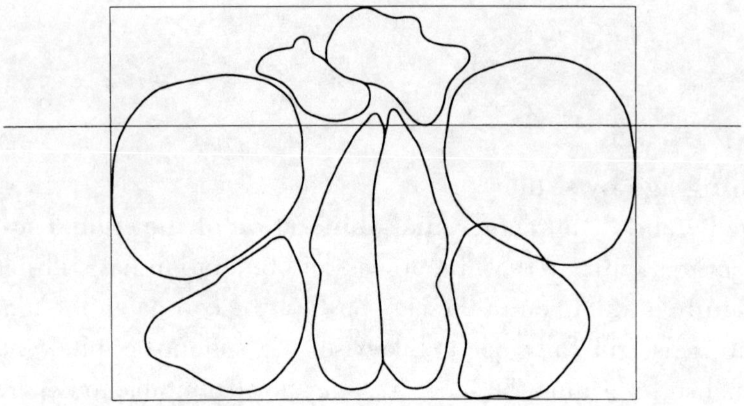

Picture 29: *Outline of the facial cavities of skull 3 from the occipito-frontal radiograph*

4.2.1.3.3. X-ray determination of the facial cavities (pict. 30)

On the occipito-frontal radiograph we can see that:

- the frontal sinuses are asymmetric, the left is larger than the right, they are leaf shaped;
- the orbits are more square-shaped. After Nemeskéri they are scored as (0);
- the maxillary sinuses are relatively symmetric, and leaf-like in shape;
- the nasal cavity is symmetric. It is shaped like a shell.

Planimetric values of the facial cavities

Skull		Frontal sinuses		Orbits		Maxillary sinuses		Nasal cavity
		Right	Left	Right	Left	Right	Left	
No. 3	S	2.76	5.08	13.09	12.91	7.30	7.93	12.20
	H	2.76	3.21	4.10	4.19	3.56	3.74	5.43
	W	1.43	2.32	3.65	3.65	2.59	2.50	3.12
	T	0.45	0.54	0.81	0.81	0.63	0.63	0.54
	Σ	7.84						
	Sh	0.45	0.63	0.89	0.89	0.63	0.81	0.54

Legend: *S = surface, H = maximal height, W = maximum width, T = W/H, Σ = total surface of frontal sinuses, Sh = shape of sinuses*

4.2.1.3.4. Paleopathological methods

The mandibular fossa is unchanged on both sides. Skull no. 3 presents no specific paleopathological features.

4.2.1.3.5. Commentary

Skull no. 3, according to the index, is classified among the short skulls. It is round in form, compact, somewhat heavier and slightly damaged. It is missing the right cheekbone and part of the orbit on the right side, both of which were damaged after death. The nasal bone is prominent and short, with no septum. According to its features this is not a typical female skull; it has larger mastoids, a moderately developed occipital protuberance and attachments for the occipital muscles. The cheekbones are also more pronounced. The shape of the orbit is between male and female, the first being more angular, the second more rounded; the dental arch is semicircular with all dental alveoli preserved.

The person had very healthy teeth, and the face was fairly broad. The remaining features indicate a female skull; glabella, inclination of the forehead, arcus superciliaris, tuber frontale and parietale.

Among specific features are the two symmetrical palatine tori next to the central suture of the palate. On the radiographs the visible hyperostosis frontalis interna continues without pathological indications; it is, however, more probable that the person suffered on account of headaches.

On the parietal bone right, paramedially behind, approx. 3 cm above the suture with the occipital bone, lies a macroscopically visible small hole which is only superficial; by using a needle we cannot pierce right through the whole thickness of the bone. It is an artefact.

The articulating surfaces of the temperomandibular joint are without special features.

Since the skull is relatively robust, there are certain features which lead us to feel that it is that of a male. Nevertheless, there are dominating features which suggest that this is actually the skull of a robustly built youngish woman. This conclusion is also confirmed, for these cases, by the significant orbit index.

THE SKULL BELONGED TO AN ADULT WOMAN, AGED 30–40 YEARS OR MORE

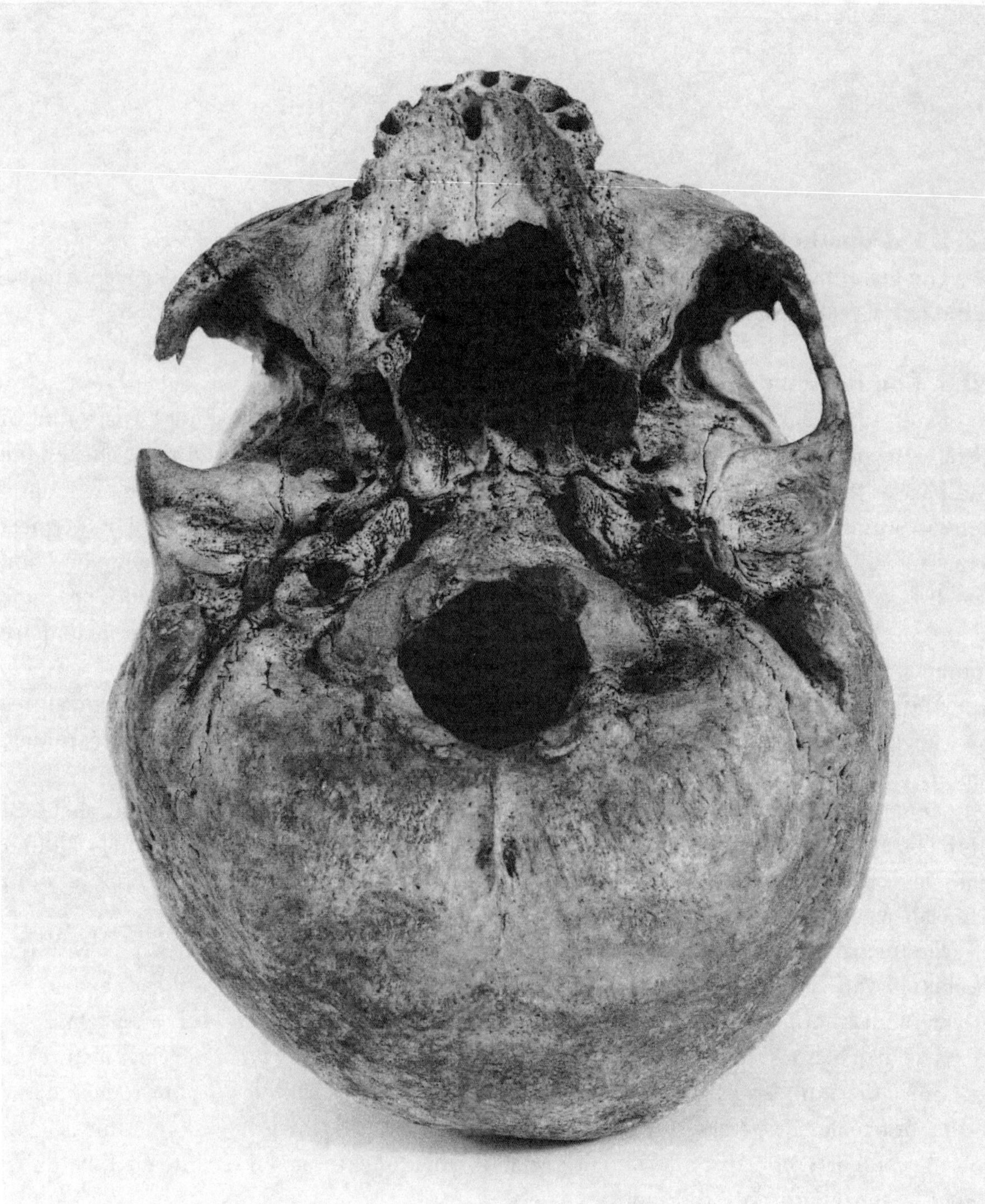

Picture 30: *Skull 4 from below reveals a change on the left occipital condyle*

4.2.1.4.

Skull 4 (pict. 30)

4.2.1.4.1. Determining gender by skull

Anthropological characters for determining gender

superciliary arches	slightly developed (–1)
glabella	unpronounced (–2)
frontal tubers	expressed (–1)
inclination of the forehead	almost vertical (–1)
supraorbital margin	unpronounced (–1)
orbital shape	more rounded (–1)
occipital relief	slightly pronounced (–2)
ext. occipital protuberance	unpronounced (–2)
mastoid processes	small, r. larger than l. (–1)
cheekbones	r. damaged, l. gracile (–2)
maxillary palatine process	ellipsoid
maxillary alveolar process	between ellipsoid and semicircular
thickness of calvaria	uniform
facial part of skull	small and narrow

Determining gender by anthropometric methods and indices

bizygomatic breadth	after reconstruction, approx. 124 mm
great occipital foramen	rounded, length 32 mmm, breadth 30 mm
mastoids	both preserved, breadth r. 30 mm, l. 22 mm
breadth of anterior nasal aperture	24 mm
cranial length	178 mm
cranial breadth	132 mm
cranial base	92 mm
cranial index	74.15 (long skull)
orbital height	r. 31.5 mm, l. 32 mm
orbital breadth	r. 35 mm, l. 35 mm
orbital index	r. 90.0, l. 91.42. Both indicate a female skull.

Determining gender by dental methods

Dental index after Pont

Breadth between first premolars	36 mm. P = 45
Breadth between first molars	37.5 mm (after reconstruction, since the alveoli of both first molars are missing). M = 79.16

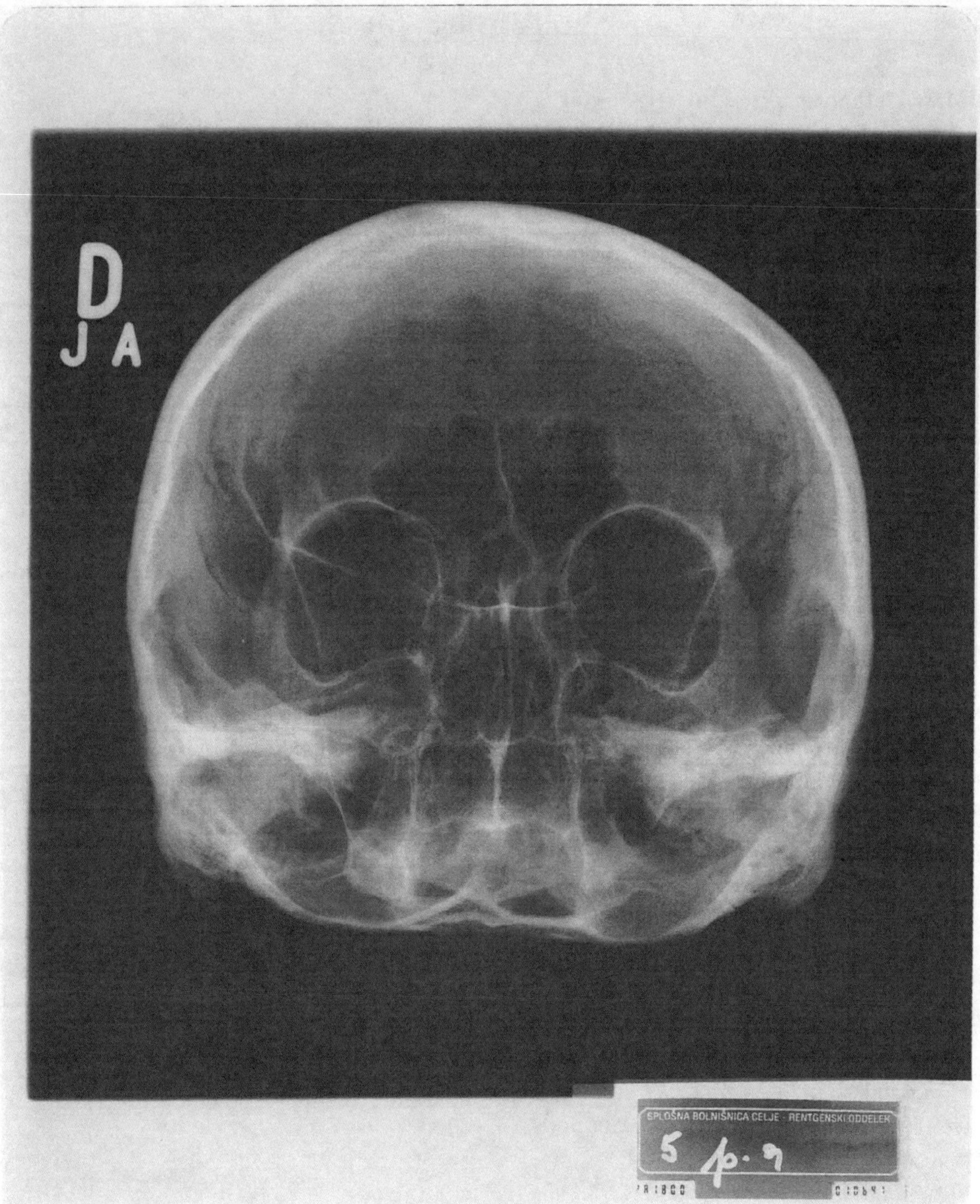

Picture 31: *Occipito-frontal radiograph of skull 4*

4.2.1.4.2 Determining age by skull

Determining age by endo- and ectocranial obliteration of the sutures

The sagittal suture is obliterated, the coronal suture distally is still partly open, while the lambdoidal is almost completely obliterated. Both the palatine medial and the spheno-occipital sutures are obliterated. The description indicates a person aged over 50 years.

Determining age by forensic-dental methods

No dentition is preserved in the maxilla. The alveolar ridge is highly atrophied, the palate is asymmetrical. On both sides the back portion of the hard palate is broken off at the level of the second molar, and the palatine bone (os palatum) is missing; the damage was probably artificially incurred.

The right dental alveoli are shallow, and are atrophic from the second premolar backwards.

On the left, the alveoli of the first four teeth are well preserved; from the second premolar backwards the maxilla is broken off.

Maxillary alveolar process from right to left:

	M3	M2	M1	PM2	PM1	C	I2	I1	I1	I2	C	PM1	PM2	M1	M2	M3
Maxilla:	a	a	a	+	+	+	+	+	+	+	+	+	+	b	b	b

Legend: *a = atrophic, b = broken off*

Determining age by X-ray methods (pict. 31)

On the antero-posterior images one may see the two orbits with rounded margins. The nasal root is broad, with the septum deviating to the left. The bone thickness frontally is relatively great, but is thinned down around the suture to the right of the bregma. On the left side view parietally one may see the prominent arterial canal. The sagittal, coronal and lambdoidal sutures are obliterated.

On the vertex right, in both the fronto-occipital and occipito-frontal radiographs, there is an oval structure approx. 1 cm large and well demarcated, with smooth margins, the so-called condensation zone, indicating either a physiological or a pathological process, which was long-lasting and which tended towards being marked off from the environment. In differential diagnostic terms, one could consider Pacchioni's canal (Grashey, Birkner, 1964), a cyst or benign growth that had sclerotized. The structure is present entirely in the calvaria between lamina int. and lamina ext.

On the radiographs of the cranial base one may see a specific feature in the pterygoid fossa: the sinus pterigoideus is exceptionally asymmetric. The other X-rays are without particular features.

no. 4: ♀, over 50 years

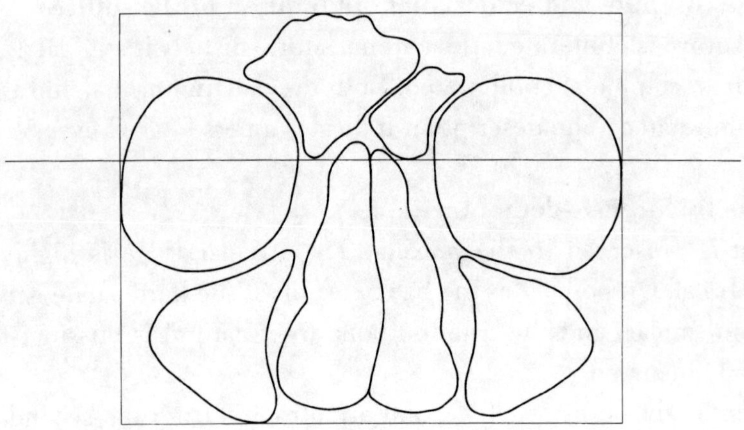

Picture 32: *Outline of the facial cavities of skull 4 from the occipito-frontal radiograph*

4.2.1.4.3. X-ray determination of the facial cavities (pict. 32)

On the occipito-frontal radiograph we can see that:

- the frontal sinuses are smaller, asymmetric, and the right is larger than the left. They are leaf shaped;
- the orbits are between angular and rounded in shape. After Nemeskéri they are scored as (–1);
- the maxillary sinuses are fairly symmetrical. They resemble a triangular shape;
- the nasal cavity is symmetrical. It is scallop shaped.

Planimetric values of the facial cavities

Skull		Frontal sinuses		Orbits		Maxillary sinuses		Nasal cavity
		Right	Left	Right	Left	Right	Left	
No. 4	S	7.84	2.50	15.76	15.58	6.06	7.84	11.75
	H	3.56	2.41	4.81	4.63	3.30	3.74	4.99
	B	2.94	1.25	3.92	3.74	2.94	2.94	3.21
	T	0.72	0.45	0.72	0.72	0.81	0.72	0.63
	Σ	10.32						
	Sh	0.72	0.63	0.85	0.81	0.63	0.63	0.63

Legend: *S = surface, H = maximal height, B = maximal breadth, T = B/H, Σ = total surface of frontal sinuses, Sh = shape of sinuses*

4.2.1.4.4. Paleopathological methods

The oval formation on the crown of the skull, which lies throughout the calvaria between lamina int. and ext. and which is described among the radiographs, may be either a physiological or a pathological growth of benign nature, like a cyst, an inflammation process – osteomyelitis or a benign growth which has sclerotized.

The calvaria reveals characters of osteoporosis. There are no other particular paleopathological features.

4.2.1.4.5. Commentary

The skull is long (on the cranial index it is relatively the longest among all 18 skulls), round in shape, rather gracile and light. It is a typical female skull, without the glabella, with deformed supraorbital arches and frontal tubers and a gently inclined forehead. The orbits are more rounded, the nasal bone longish and gracile; only the back part of the nasal septum is preserved, and is deviated to the left. The left cheekbone is feminine and intact, the right is missing the arch. The mastoids are moderately large. Occipitally, the muscle attachments and the occipital protuberance are slightly pronounced. The great occipital foramen is rounded, and on the dorsal side the left occipital condyle is altered.

The person had a high forehead, slightly expressed supra-orbital arches, a narrow nasal root, narrowish cheekbones and an elliptically shaped palate and dental arch, all of which together gave her face a narrow appearance. The dental alveoli are more poorly preserved on the right than on the left. This indicates that the person had lost a considerable number of teeth during her lifetime. All the cranial sutures are obliterated; they indicate an older person's skull.

The cranial bones are thinned down due to ageing porosis; the occipito-frontal radiograph reveals in the bone at the crown of the skull an oval formation, similar to a cyst or a bone inflammation – osteomyelitis, or a benign growth which had sclerotized.

THE SKULL BELONGED TO AN OLDER WOMAN, OVER THE AGE OF 50

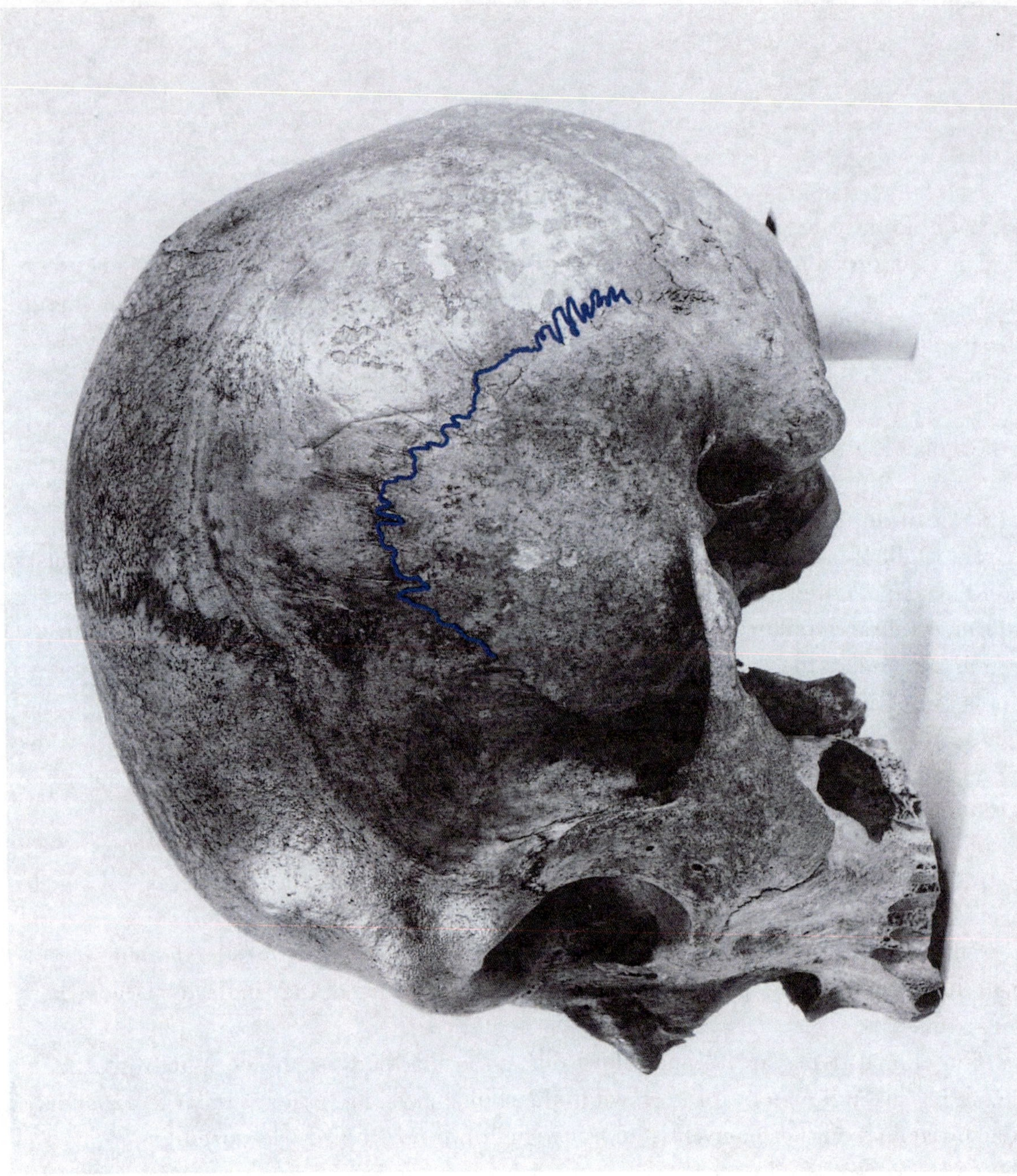

Picture 33: *Skull 5 from the side. Of interest is the high temporal bone, similar to that in skulls 6 and 7*

4.2.1.5.
Skull 5 (pict. 33)

4.2.1.5.1. Determining gender by skull

Anthropological characters for determining gender

superciliary arches	strongly developed (+2)
glabella	pronounced (+2)
frontal tubers	moderately pronounced (+1)
inclination of the forehead	steeply inclined (+2)
supraorbital margin	rounded
orbital shape	rectangular (+2)
occipital relief	prominent (+2)
ext. occipital protuberance	prominent (+2)
mastoid processes	prominent (+2)
cheekbones	moderately prominent (+l)
maxillary palatine process	semicircular
maxillary alveolar process	semicircular
thickness of calvaria	occipitally larger than frontally
facial part of skull	pronounced

Determining gender by anthropometric methods and indices

bizygomatic breadth	135 mm
great occipital foramen	round, length 34 mm, breadth 32 mm
mastoids	both preserved, breadth r. 32 mm, l. 25 mm
breadth of anterior nasal aperture	27 mm
cranial length	175 mm
cranial breadth	145 mm
cranial base	102 mm
cranial index	82.85 (short skull)
orbital height	r. 31.5 mm, l. 32 mm
orbital breadth	r. 36 mm, l. 36 mm
orbital index	l. 88.88, r. 87.50 (male feature)

Determining gender by dental methods

Dental index after Pont:

Breadth between first premolars	32 mm. P = 40
Breadth between first molars	47 mm. M = 80

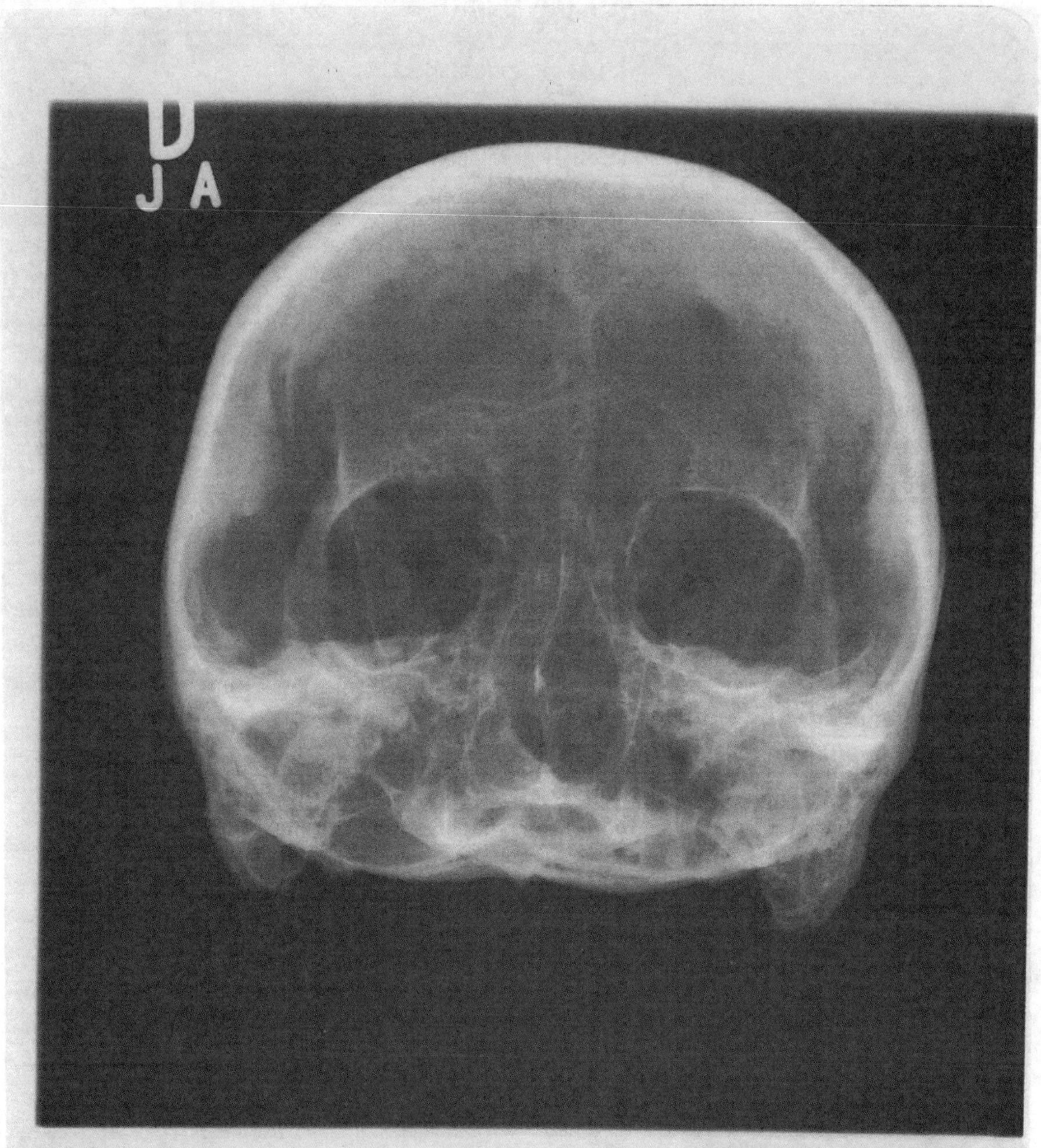

Picture 34: *Occipito-frontal radiograph of skull 5*

4.2.1.5.2. Determining age by skull

Determining age by endo- and ectocranial obliteration of the sutures

The sagittal suture has obliterated; still apparent in the coronal suture is a portion of the pars complicata, while the lambdoidal still retains an open asterical part. The sphenooccipital suture has already obliterated as well, but the medial palatine not yet.

The above description of the sutures indicates a person aged around 50 years.

Determining age by forensic-dental methods

Maxilla: the alveolar process is without teeth. The left palatine tuber is broken off (artefact). The palatine torus is moderately pronounced, and the incisive foramen is large.

The right front dental alveoli are poorly preserved, the back ones better; the alveoli of the second incisor and canine are only partly preserved.

The alveoli of the first and second incisors left are preserved. Part of the alveolar ridge, where one would expect to find the canine and both premolars, is broken off. The alveoli of all three molars are only partially preserved.

Maxillary alveolar process from right to left:

	M3	M2	M1	PM2	PM1	C	I2	I1	I1	I2	C	PM1	PM2	M1	M2	M3
Maxilla:	+	+	+	+	+	+p	+p	+	+	+	+b	+b	+b	+p	+p	+p

Legend: *p = partially preserved, b = broken off*

Determining age by X-ray methods (pict. 34)

On the fronto-occipital radiograph one sees a large male skull. The orbits are more angular, the nasal bone is broad, and only the back part of the nasal septum, which is straight, is preserved. Frontally visible are the unpronounced Pacchionian bodies. The mastoids are of different shape, and also differ in structure. The structure of the septa in the left mastoid is considerably rarer than in the right. There exists the possibility of subacute, or chronic, mastoiditis in adult age. With an image of the mastoids after Schiller one might be able to learn more about this pathology. The lateral images reveal a prominent sulcus of the meningeal artery. Likewise, the side and base images indicate that the occipital bone is thicker than the frontal. The cranial sutures are obliterated. Otherwise, there are no specific features in the images.

no. 5: ♂, aged around 50

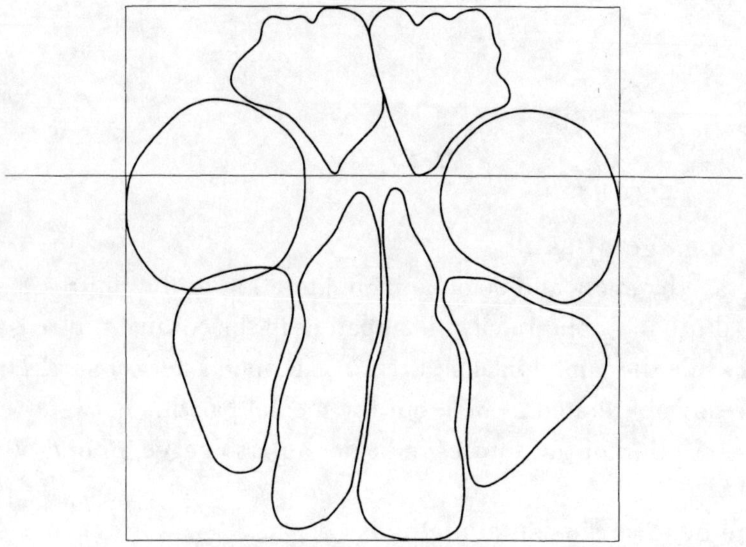

Picture 35: *Outline of the facial cavities of skull 5 from the occipito-frontal radiograph*

4.2.1.5.3. X-ray determination of the facial cavities (pict. 35)

The occipito-frontal radiograph shows that:

- the frontal sinuses are slightly asymmetric and are fanlike-pyramidal in shape;
- the orbits are more rectangular in shape. After Nemeskéri, they are scored as (+2);
- the maxillary sinuses are relatively symmetrical. They resemble a triangular shape;
- the nasal cavity is long and relatively symmetrical. It is scallop shaped with serrated margins.

Planimetric values of the facial cavities

Skull		Frontal sinuses		Orbits		Maxillary sinuses		Nasal cavity
		Right	Left	Right	Left	Right	Left	
No. 5	S	9.08	9.62	17.54	17.89	8.46	7.75	18.96
	H	3.83	4.19	4.72	4.81	4.37	4.90	6.41
	B	3.30	2.94	4.30	4.37	2.94	2.50	3.74
	T	0.81	0.63	0.81	0.81	0.63	0.45	0.54
	Σ	18.69						
	Sh	0.63	0.63	0.81	0.81	0.54	0.54	0.63

Legend: *S = surface, H = maximal height, B = maximal breadth, T = B/H, Σ = total surface of frontal sinuses, Sh = shape of sinuses*

4.2.1.5.4 Paleopathological methods

The maxillary joints show no visible degenerative changes. The altered left mastoid indicates the pathology, most probably a subchronic or chronic inflammatory process, which had fistulated. Clinically, the person may have had weaker hearing in the left ear.

4.2.1.5.5. Commentary

According to absolute values and to the cranial index, this skull is short. It is of round shape, massive and slightly damaged. It bears typical male characters: the glabella is pronounced, so is the superciliary arch. On the right frontal bone it has an oblique bone groove. The orbits are more rectangular, with prominent masculine features according to the orbital index. The nasal root is broad, the nasal bone sharp and missing the front part of the nasal septum, while the back part is preserved and straight. The mastoids are massive and asymmetric; the left one has a protuberance at the back. The great occipital foramen is round, the occipital protuberance is highly pronounced, and the occipital condyles are also large. Only the frontal incline and the frontal tubers are not too prominent and the cheekbones are not highly pronounced.

The palatine part of the maxilla is semicircular, as is the alveolar, which is also confirmed by the dental index after Pont. All the recorded features, together with Campers' angle, indicate a face of average breadth with moderately pronounced cheekbones.

The altered left mastoid is indicative of an inflammatory process which has fistulated. The person probably had weaker hearing in that ear. To the right, on the frontal bone, there are two bone outgrowths (exostoses) on the skull, 4 and 6 mm in size, respectively, and 1 mm in height.

THE SKULL BELONGS TO AN ADULT MALE, AGED ABOUT 50 YEARS

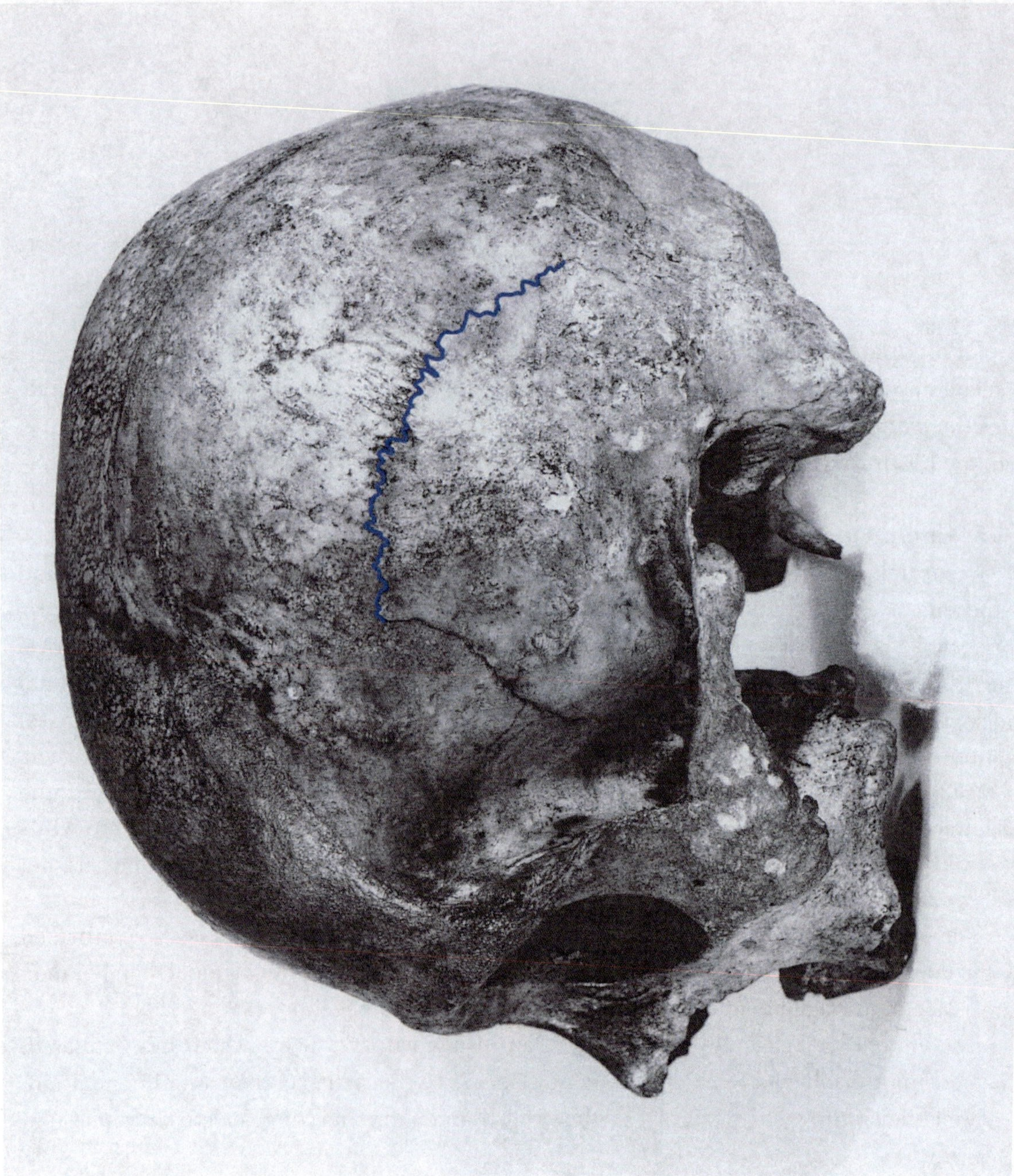

Picture 36: *Skull 6, side view. It has the typical high temporal bone, also found in skulls 5 and 7*

4.2.1.6.
Skull 6 (pict. 36)

4.2.1.6.1. Determining gender by skull

Anthropological characters for determining gender

superciliary arches	strongly developed (+2)
glabella	highly pronounced (+2)
frontal tubers	strongly pronounced (+2)
inclination of the forehead	steeply inclined (+2)
supraorbital arch	highly pronounced (+2)
orbital shape	trapezoid (+2)
	moderately pronounced (+1)
occipital relief	highly prominent (+2)
ext. occipital protuberance	highly prominent (+2)
mastoid processes	highly pronounced (+2)
cheekbones	highly pronounced (+2)
maxillary palatine process	semicircular
maxillary alveolar process	semicircular
thickness of calvaria	occipitally greater than frontally
facial part of skull	pronounced

Determining gender by anthropometric methods and indices

bizygomatic breadth	135 mm
great occipital foramen	rounded, length 33 mm, breadth 30 mm
mastoids	both preserved, breadth r. 34 mm, l. 30 mm
breadth of anterior nasal aperture	27 mm
cranial length	176 mm
cranial breadth	140 mm
cranial base	103 mm
Campers' angle	88 degrees
cranial index	79.54 (medium length skull)
orbital height	r. 35 mm, l. 35 mm
orbital breadth	r. 40 mm, l. 40 mm
orbit index	87.5 (male feature)

Determining gender by dental methods

Dental index after Pont

Breadth between first premolars	37 mm. P = 46.3
Breadth between first molars	47 mm. M = 78.33

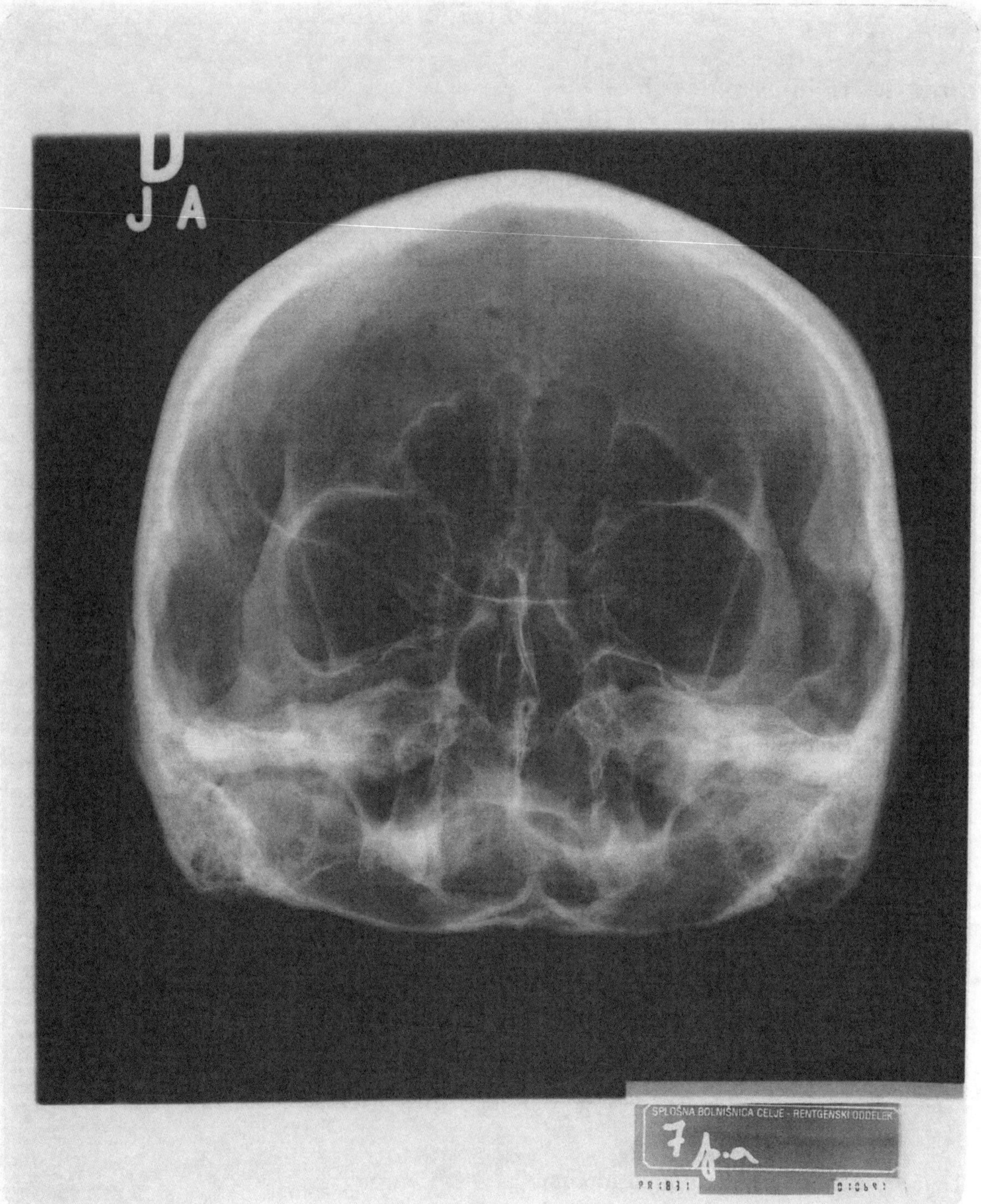

Picture 37: *Occipito-frontal radiograph of skull 6*

4.2.1.6.2. Determining age by skull

Determining age by endo- and ectocranial obliteration of the sutures

The sagittal and coronal sutures are considerably obliterated; the asteric part of the lambdoidal suture is still not obliterated. The facial sutures and the medial palatine suture are also obliterated.

Determining age by forensic-dental methods

Maxilla: in the toothless alveolar ridge, only some of the dental alveoli are preserved. Many of them are atrophic, particularly on the left side, where almost complete atrophy has occurred. The person had clearly lost the premolars and molars at least two decades before death. On the right side the alveolar ridge from the second molar has been distally broken off. The incisive foramen is pronounced, but the palatine torus is not.

On the right, the alveoli of the first incisor, the first premolar, and the first and second molar are preserved. The intermediary missing alveoli of the second incisor, the canine and the second premolar are ossified.

On the left side, the alveoli of both incisors and the canine are preserved, the others are ossified.

Maxillary alveolar process from right towards left:

	M3	M2	M1	PM2	PM1	C	I2	I1	I1	I2	C	PM1	PM2	M1	M2	M3
Maxilla:	–	+	+	t	+	t	t	+	+	+	+	t	t	t	t	t

Legend: *t = tooth*

Determining age by X-ray methods (pict. 37)

On the fronto-occipital radiograph one can see a larger skull with trapezoid orbits, and a broad nasal bone with the septum deviating considerably to the right. The bone is thicker occipitally than frontally.

Frontally, parietally right and occipitally there are numerous stains of irregular shapes. The formation dorsoparietally left is larger and without sharp margins. On the basis of the macroscopic examination of the skull, the X-ray analysis, and the assessment that the person had lived to the age of 50 or more, it was concluded that this was most probably a formation of physiological Pacchionian bodies.

On the side view right it can be seen that the strongly pronounced sulcus of the diploic vein projects into the course of the coronal suture (Grashey, Birkner 1964). Further analysis with the aid of CT images indisputably confirmed this supposition.

The other radiographs had no specific features.

no. 6: ♂, over 50 years

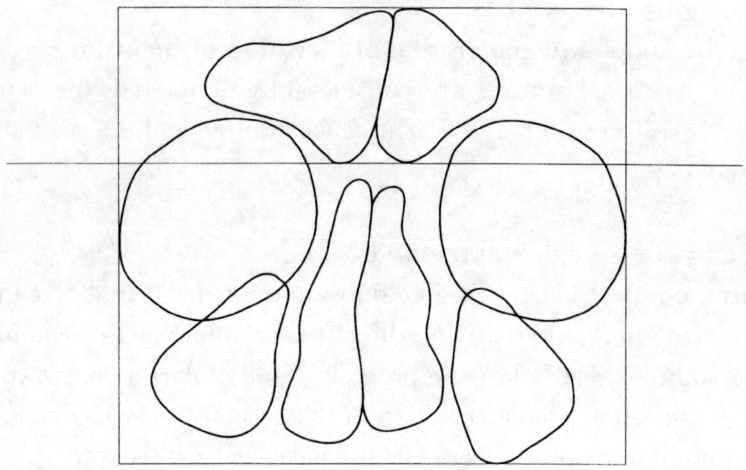

Picture 38: *Outline of the facial cavities of skull 6 from the occipito-frontal radiograph*

4.2.1.6.3. X-ray determination of the facial cavities (pict. 38)

On the occipito-frontal radiograph one may see that:

- the frontal sinuses are asymmetrical, the right is larger than the left, and they are fan shaped;
- the orbits are trapezoid in shape. After Nemeskéri they are scored as (+2);
- the maxillary sinuses are fairly symmetrical. They resemble a triangle in shape;
- the nasal cavity is symmetrical. It is scallop shaped.

Planimetric values of the facial cavities

Skull		**Frontal sinuses**		**Orbits**		**Maxillary sinuses**		**Nasal cavity**
		Right	**Left**	**Right**	**Left**	**Right**	**Left**	
No. 6	S	12.64	7.48	18.87	19.58	6.41	6.95	15.31
	H	4.99	3.65	5.17	4.99	3.74	4.01	5.79
	B	4.01	2.67	4.30	4.54	2.41	2.50	3.65
	T	0.72	0.63	0.72	0.81	0.54	0.54	0.54
	Σ	20.12						
	Sh	0.63	0.72	0.89	0.89	0.63	0.54	0.63

Legend: *S = surface, H = maximal height, B = maximal breadth, T = B/H, Σ = total surface of frontal sinuses, Sh = shape of sinuses*

4.2.1.6.4. Paleopathological methods

Both mandibular fossae indicate degenerative changes.

Skull no. 6 has an interesting pathology. Suspicions were raised by the formations, ranging in size from 2 to 10 mm, distributed over the frontal, parietal and occipital surface of the skull, uneven in shape and different in colour from the surrounding bone. From the radiographs these numerous diffuse thinned structures were reminiscent of multiple myeloma or of an osteolytic process. The macroscopic appearance of the skull, however, adjusted to the X-ray findings, indicated rather a benign bone formation – exostosis or benign osteoma.

The CT images revealed that parietally on the right the sulcus of the diploic vein was strongly pronounced, which is also one of the indirect indicators for a mature person.

4.2.1.6.5. Commentary

According to the measurements and the cranial index, the skull is of average length, longer than most of the others, round in shape, massive and almost undamaged. By its features, it is a typical male skull. The person had a pronounced glabella, superciliary arches and frontal tuber, and an inclined forehead. The occipital relief and the muscle attachments are powerful, and the mastoid prominent.

The orbits are trapezoid shape, and their indices typically male. The nasal root is broad, the septum curved towards the right. The cheekbones are moderately pronounced. In the semicircular dental arch only a few alveoli are preserved – they are mainly atrophic, since during his life the person had lost several teeth, though he had preserved the upper incisors left and right until the end of his life. The dental index after Pont indicates an alveolar arch of average breadth, while Campers' angle points to an orthognathic maxilla. The bone traits described above indicate a non-robust face of average breadth, with more pronounced supraorbital arches and a sturdy nasal bone.

To the left, on the frontal and parietal bones, benign osseous formations can be observed: size 2–10 mm, exostoses or benign osteomas. The radiograph indicates a diffuse thinned osseous structure, as in the case of osteolytic processes or multiple myeloma. Since it was not possible to explain the features described by radiographs, tomographic images were taken. These revealed that they were only strongly pronounced physiological Pacchionian bodies frontally, right parietally and occipitally, while on the side views left there is a highly prominent sulcus of the diploic vein.

The mandibular fossae indicate degenerative changes.

The record above is indicative of the skull of an elderly male. It does not allow for conclusions to be made about the cause(s) of death.

THE SKULL BELONGED TO AN ADULT MALE, AGED OVER 50 YEARS

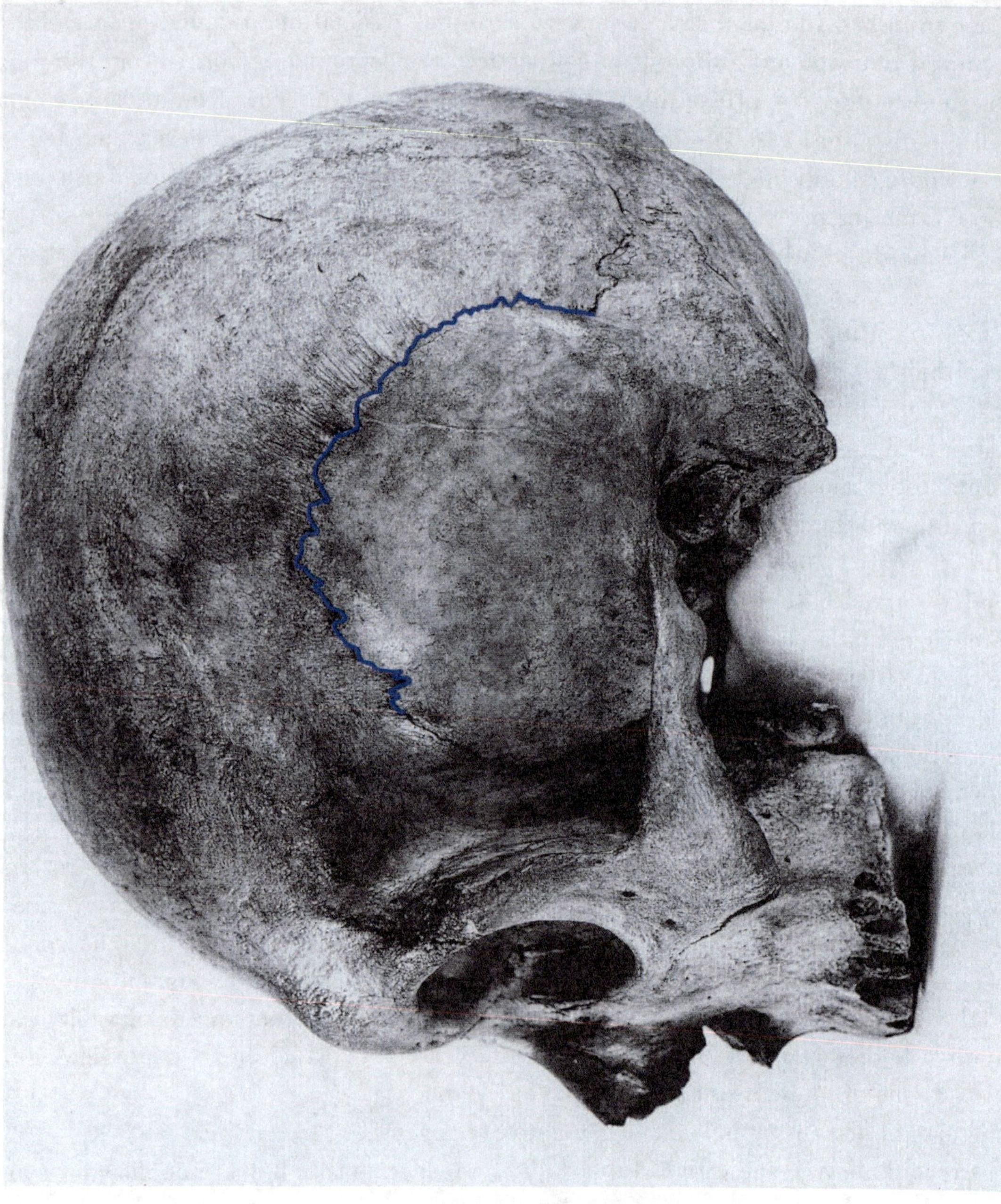

Picture 39: *Skull 7, side view. It has a high temporal bone, similar to those of skulls 5 and 6*

4.2.1.7.
Skull 7 (pict. 39)

4.2.1.7.1. Determining gender by skull

Anthropological characters for determining gender

superciliary arches	strongly developed (+2)
glabella	strong (+2)
frontal tubers	strongly expressed (+2)
inclination of the forehead	steeply inclined (+2)
supraorbital margin	highly expressed (+2)
orbital shape	square (+1)
occipital relief	pronounced (+2)
ext. occipital protuberance	moderately pronounced (+l)
mastoid processes	strong (+2)
cheekbones	moderately expressed (+1)
maxillary palatine process	semicircular
maxillary alveolar process	semicircular
thickness of calvaria	occipitally greater than frontally
facial part of skull	pronounced

Determining gender by anthropometric methods and indices

bizygomatic breadth	140 mm
great occipital foramen	round, length 37 mm, breadth 32 mm
mastoids	both preserved, breadth l. 42 mm, r. 26 mm
breadth of anterior nasal aperture	27 mm
cranial length	181 mm
cranial breadth	146 mm
cranial base	102 mm
Camper's angle	90 degrees
cranial index	80.66 (medium length skull)
orbital height	r. 31 mm, l. 33 mm
orbital breadth	r. 38 mm, l. 38 mm
orbital index	r. 81.57, l. 86.84. Indices typical of a male skull.

Determination of gender by dental methods

Dental index after Pont:

Breadth between first premolars	37 mm. P = 46.3
Breadth between first molars	47 mm. M = 78.33

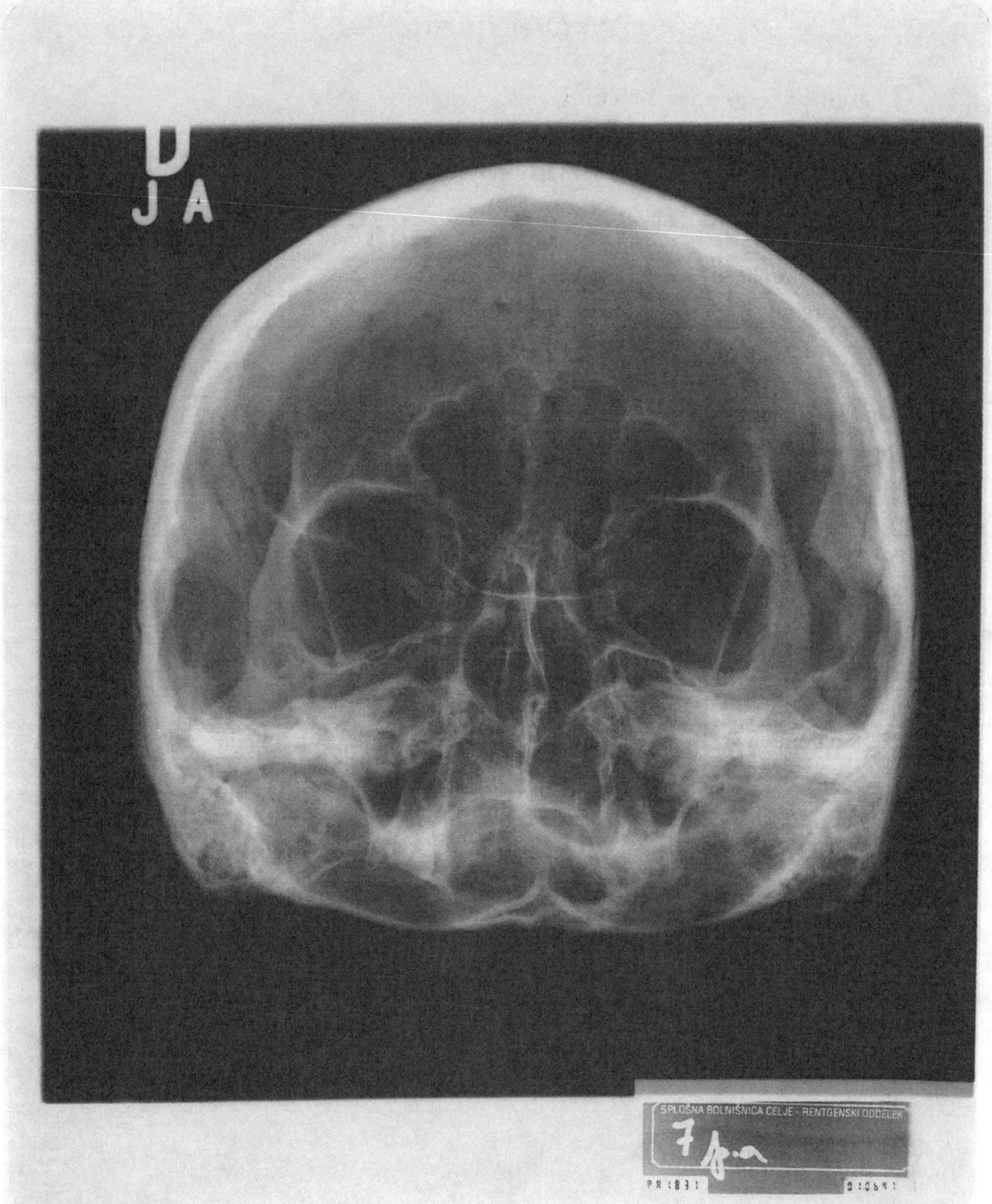

Picture 40: *Occipito-frontal radiograph of skull 7*

4.2.1.7.2. Determining age by skull

Determining age by endo- and ectocranial obliteration of the sutures

Visible are the remains of the non-obliterated sagittal suture on the crown of the skull and of the coronal suture on both sides on the temporal part. The spheno-occipital suture is obliterated, but not yet the medial palatine. The facial sutures are still not closed as well. The description indicates an adult person, aged about 30 years or more.

Determining age by forensic-dental methods

The maxilla is without teeth, although the alveoli of all 16 teeth are present. The person clearly had very healthy teeth, since they did not fall out until after his death. The alveolus of the first molar left is open into the maxillary sinus. The medial palatine suture is still not closed; on the right pharyngeal part the palatine torus is slightly developed. The incisive foramen is moderately pronounced.

Maxillary alveolar process from right towards left:

	M3	M2	M1	PM2	PM1	C	I2	I1	I1	I2	C	PM1	PM2	M1	M2	M3
Maxilla:	+	+	+	+	+	+	+	+	+	+	+	+	+	+	+	+

Legend: *+ = alveolus present*

Determining age by X-ray methods (pict. 40)

The fronto-occipital view shows a massive skull with rectangular orbits, and a strong long nasal bone, with the nasal septum deviating slightly to the left. The Pacchionian bodies are within the physiological boundaries. The incompletely obliterated sagittal suture is also clearly visible. On the vertex, the skull is thickened in the sense of frontal internal hyperostosis (Grashey, Birkner, 1964), which may be a physiological phenomenon.

The sphenoidal sinus is hypertrophic. The radiographs showed no other specific features.

no. 7: ♂, aged 30–40

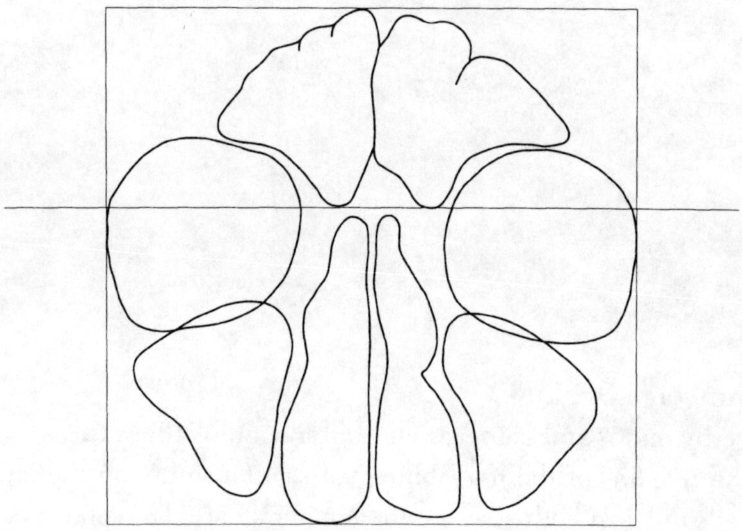

Picture 41: *Outline of the facial cavities of skull 7 from the occipito-frontal radiograph*

4.2.1.7.3. X-ray determination of the facial cavities (pict. 41)

The occipito-frontal radiograph shows that:

- the frontal sinuses are large, almost symmetrical, and are fan-pyramid shaped;
- the orbits are more rectangular shape. After Nemeskéri they are scored as (+1);
- the maxillary sinuses are relatively symmetrical. They resemble a triangular shape;
- the nasal cavity is relatively symmetrical, scallop shaped with serrated margins.

Planimetric values of the facial cavities

Skull		Frontal sinuses		Orbits		Maxillary sinuses		Nasal cavity
		Right	Left	Right	Left	Right	Left	
No. 7	S	11.84	13.62	18.43	18.25	9.26	6.23	17.09
	H	5.08	5.08	4.90	4.90	4.01	3.56	6.15
	B	3.39	4.01	4.45	4.30	3.12	2.50	3.56
	T	0.63	0.72	0.81	0.81	0.72	0.63	0.54
	Σ	25.46						
	Sh	0.54	0.54	0.89	0.89	0.72	0.72	0.63

Legend: *S = surface, H = maximal height, B = maximal breadth, T = B/V, Σ = total surface of frontal sinuses, Sh = shape of sinuses*

4.2.1.7.4. Paleopathological methods

The mandibular fossae are smooth, without degenerative changes. Although the mastoids are of different largeness, they do not reveal pathological changes. Careful endoscopic examination was also made of the left maxillary sinus, which is probably artificially linked to the left first molar, but is without specific traits. The suspicion of an inflamed process (sinusitis) was excluded.

4.2.1.7.5. Commentary

According to the cranial index, this skull is of medium length. It is round shape, massive and almost undamaged. It reveals most of the characters typical of a male skull: particularly pronounced are the glabella, the supraorbital arches and the frontal tuber. The forehead is quite steeply inclined. On the occiput the relief is likewise pronounced, which indirectly indicates strong occipital muscles. The mastoids left and right are asymmetric, though without characters of inflammation. The occipital protuberance is somewhat smaller. The orbits are rectangular in shape, and the orbital indices display male features. The nasal bone is wide, with the nasal septum deviating to the left. Campers' angle indicates an orthognathic maxilla. The dental index after Pont reveals a regular semicircular shape of the dental arch of the maxilla, mainly preserved dental alveoli, and healthy teeth which fell out posthumously. The above described features are indicative of a harmoniously structured, fairly broad upper part of the face with a relatively high forehead, broad cheekbones and a sturdy nasal bone. On account of the deviation of the nasal septum, the person most probably breathed with some difficulty through the nose.

The mandibular fossae are without changes. The left sinus of the maxilla is joined by a fistula to the first left molar, although there are no signs of sinusitis. The frontal bone is thinned down left of the bregma. Above the right lacrimal sulcus there extrudes a smallish exostosis.

The features described, which determine gender, undoubtedly indicate a skull of male gender; while the traits which determine age point to a mature adult male.

THE SKULL BELONGED TO AN ADULT MALE, AGED 30–40 YEARS

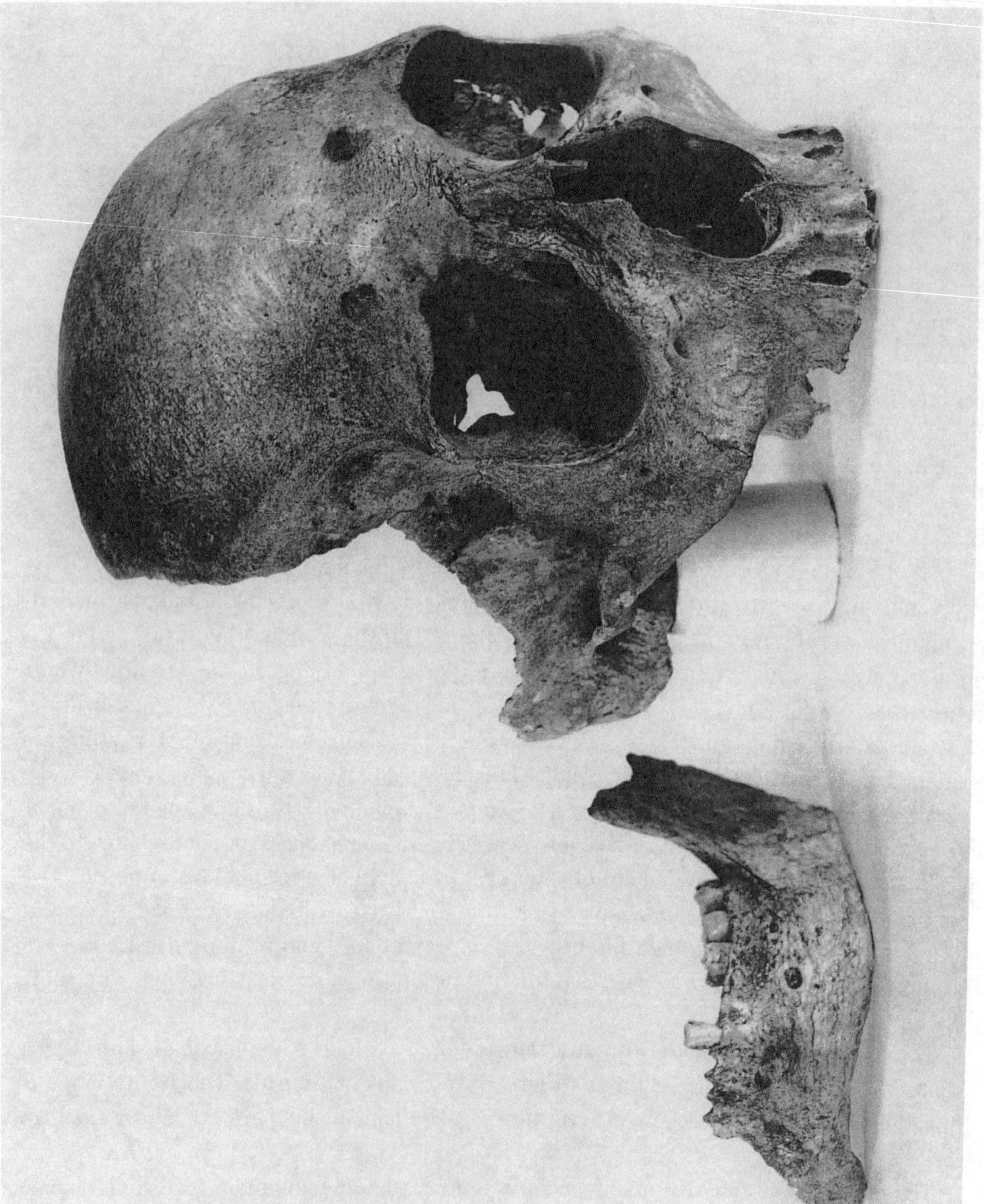

Picture 42: *Skull 8 from semi-profile, showing: severe injury, artificial stains on the skull and mandible, of unknown origin, presented together with the skull*

4.2.1.8.
Skull 8 (pict. 42)

4.2.1.8.1. Determining gender by skull

Anthropological characters for determining gender

superciliary arches	strongly developed (+2)
glabella	strong (+2)
frontal tubers	strongly pronounced (+2)
inclination of the forehead	moderately steep (+l)
supraorbital margin	moderately expressed (+1)
orbital shape	rectangular (+1)
occipital relief	broken off
ext. occipital protuberance	broken off
mastoid processes	broken off
cheekbones	highly prominent (+2)
maxillary palatine process	rounded
maxillary alveolar process	semicircular
thickness of calvaria	frontally the skull is relatively thick
facial part of skull	pronounced

Determining gender by anthropometric methods and indices

bizygomatic breadth	137 mm
great occipital foramen	broken off
mastoids	both broken off
breadth of anterior nasal aperture	30 mm
cranial length	undeterminable, broken off
cranial breadth	undeterminable, broken off
cranial base	after reconstruction, approx. 100 mm
cranial index	undeterminable
orbital height	r. 38 mm, l. 38 mm
orbital breadth	r. 40 mm, l. 40 mm
orbital index	92.68

Determining gender by dental methods

Dental index after Pont	
Breadth between first premolars	42 mm. P = 52.5
Breadth between first molars	53 mm. M = 88.33

4.2.1.8.2. Determining age by skull

Determining age by endo- and ectocranial obliteration of the sutures

The sagittal suture, as much of it is preserved, is obliterated. So too is the coronal suture; the lambdoidal however, is not preserved. The medial palatine suture is open. The facial sutures are also still open.

The above description indicates an adult person, aged around 30 years.

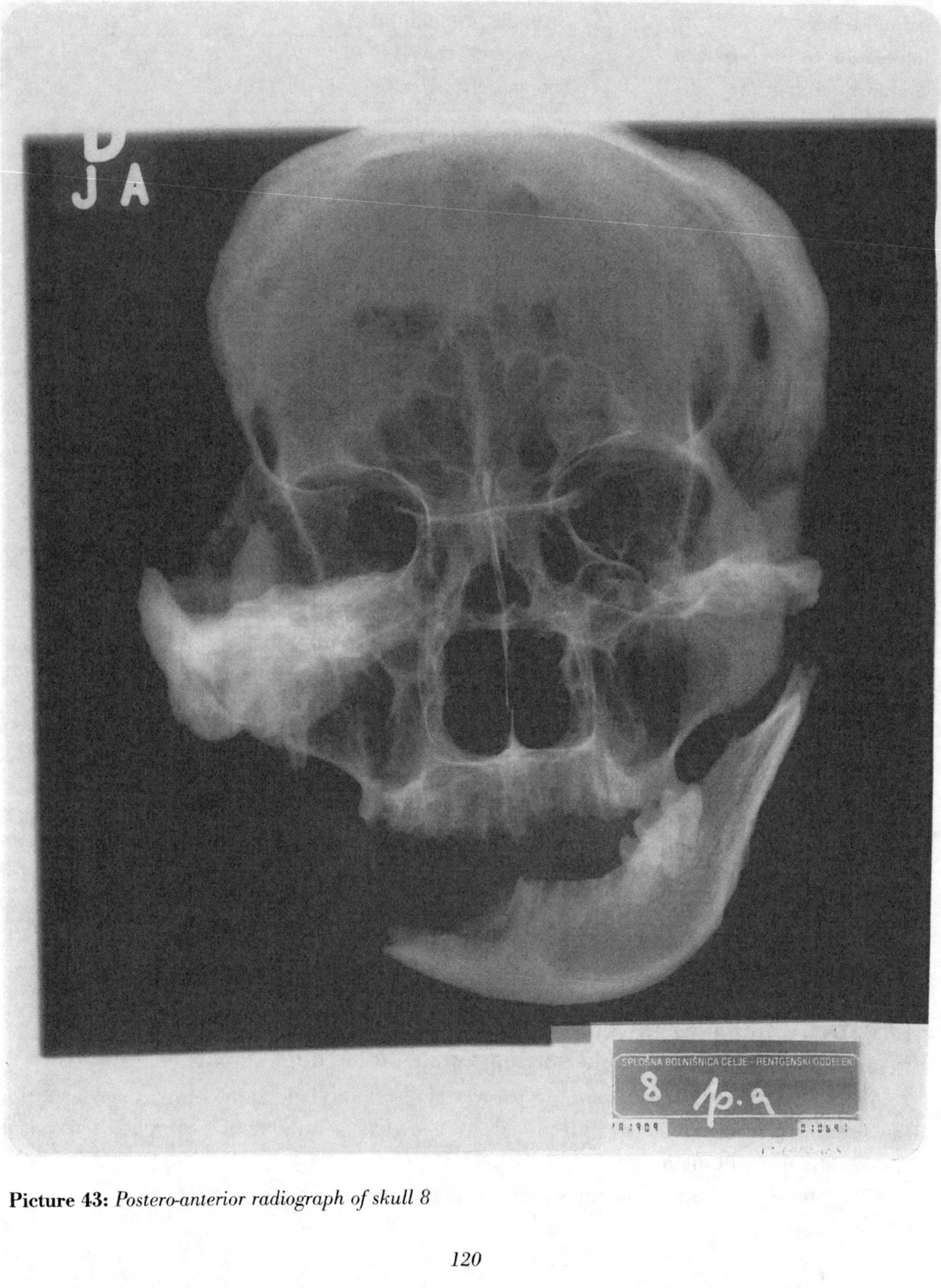

Picture 43: *Postero-anterior radiograph of skull 8*

Determining age by forensic-dental methods

The maxilla retains no preserved teeth, however there are dental alveoli left and right up to the second molar. To the left, by the third molar, part of the maxillary tuber is broken off.

On the right, all alveoli, with the exception of the first molar, indicate that the person retained his teeth until death; the first molar, however, was lost during lifetime, since the alveolus is atrophic, with rounded margins on the palatine side, yet the depth of the alveolus is still only 3 mm. The third molar – the wisdom tooth – had possibly not developed (hypodontia).

On the left side of the maxillary alveolar ridge, all the alveoli – apart from the second premolar (atrophy) – are deep, and indicate that the teeth fell out after death. No conclusion can be drawn regarding the third left molar, since the maxillary tuber is broken off. Next to the medial palatine suture the palatine torus is paramedially expressed.

Maxillary alveolar process from right towards left:

	M3	M2	M1	PM2	PM1	C	I2	I1	I1	I2	C	PM1	PM2	M1	M2	M3
Maxilla:	h	+	a	+	+	+	+	+	+	+	+	+	a	+	+	b

Legend: *a = atrophic, h = hypodontia, b = broken off*

Mandible: it cannot be concluded with certainty that the only preserved mandible belonged to skull 8. This is how it was preserved in the Regional Museum (Pokrajinski muzej) in Celje. Since it cannot be proved to which skull the mandible belonged, the gender and age of skull 8 will be determined without considering the conclusions offered by this mandible.

Only the left half – the body and branch of the mandible – is preserved; the condilar processus and coronid processus, and the right part of the body are missing. On the left side 4 teeth are preserved: the canine and all three molars. The alveoli of the second incisor and the first and second premolars are empty. The first incisor has a preserved root. In the preserved right branch of the mandible the alveoli of both incisors are empty.

The mandible is of medium construction and measures in breadth: left, in the region of the preserved second molar 18 mm, of the first molar 14 mm, and of the canine 15 mm. The entrance into the mandibular canal, the mandibular lingula and the mental foramen between the premolars are preserved. The mandibular angle is 1200. The first incisor has a root in the alveolus. The first molar is the most abraded, and the abrasion has already reached the dentine. The second molar has abraded occlusion and labial surface, so that the pale yellowish enamel shines through. The third molar is abraded at the level of the dentine and is without caries.

Determining age by X-ray methods (pict. 43)

The fronto-occipital radiographs show a partially preserved skull. The orbits are rectangular; the nose is without specific traits, with a preserved straight septum. Frontally, shadows are present which are not Pacchionian bodies; they are reminiscent of metastases. They are attributed to the macroscopically visible stains on the forehead of the skull.

On skull 8 – considering the numerous missing parts no other X-ray particularities are observed.

no. 8: ♂, aged around 30 years

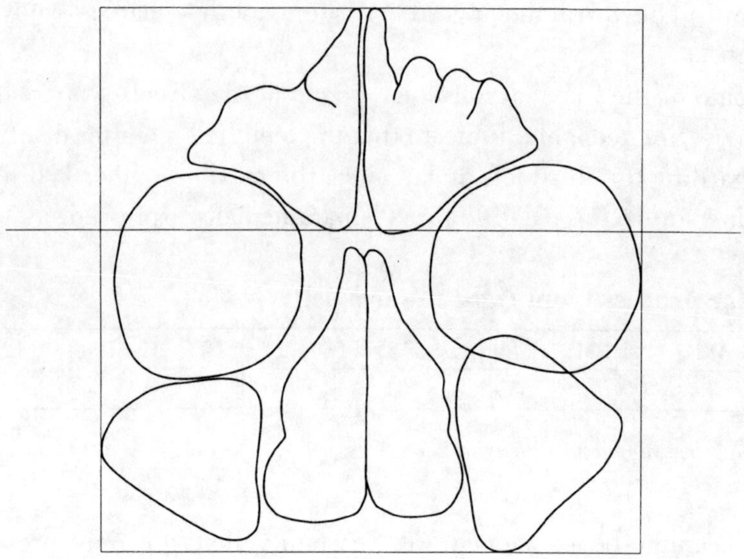

Picture 44: *Outline of the facial cavities of skull 8 from the occipito-frontal radiograph*

4.2.1.8.3. X-ray determination of the facial cavities (pict. 44)

The occipito-frontal radiograph shows that:

- the frontal sinuses are large, slightly asymmetric and pyramid shaped;
- the orbits are rectangular. After Nemeskéri they are scored as (+l);
- the maxillary sinuses are also slightly asymmetric. They resemble the shape of a triangle;
- the nasal cavity is symmetrical. It is scallop shaped.

Planimetric values of the facial cavities

Skull		Frontal sinuses		Orbits		Maxillary sinuses		Nasal cavity
		Right	Left	Right	Left	Right	Left	
No. 8	S	8.08	10.72	16.32	16.32	14.40	9.76	18,80
	H	5,04	4.72	4,56	4.48	4.48	4.00	6.24
	B	2.32	3.60	3.84	3.92	3.84	2.64	3.60
	T	0.40	0,64	0.72	0.72	0.72	0.56	0.48
	Σ	18.80						
	Sh	0,40	0.32	0.78	0.72	0,64	0.72	0.56

Legend: *S = surface, H = maximal height, B = maximal breadth, T = B/H, Σ = total surface of frontal sinuses, S = shape of sinuses*

4.2.1.8.4. Paleopathological methods

The articulating surface of the temporomandibular joint on the left side displays no degenerative changes. The right mandibular fossa is broken off.

In the case of skull 8, from the paleopathological aspect, it was important to explain the unusual shadows frontally on the fronto-occipital radiograph. The suspicion of a metastatic carcinoma was refuted by comparison with the macroscopically visible structures on the skull forehead, which do not accord with the X-ray shadows.

4.2.1.8.5. Commentary

The skull was seriously damaged posthumously, and consequently it was not possible to verify all the desired parameters. Following reconstruction and calculation of the cranial index, it is a skull of medium length. The facial part is mainly preserved, and therefore one may draw most of the conclusions about it. Partially preserved are the parietal part and part of the cranial vault. On both sides, most of the temporal and all of the occipital bones are missing. On the facial part, in particular the pronounced male features are visible: expressed glabella, prominent supraorbital arches, expressed frontal tuber, robust cheekbones. The forehead is moderately steep, the shape of the orbits rectangular. The latter are typical of the male gender, as is confirmed by the orbital index. The nasal root is broad, the nasal bone partly broken off, the septum is straight. On the skull surface above the glabella and in the centre above the right orbit there are two stains of dark colour and round shape, the size of a 1 euro coin (7 mm). They are artefacts. Campers' angle indicates an orthognathic maxilla and regularly positioned teeth. The dental index after Pont indicates a semicircular broadish dental arch and, consequently, a rather broad face with pronounced cheekbones. During his lifetime the person had retained most of the teeth.

On the skull there are no signs of illness, and also the temporomandibular articulating surfaces are not abraded. The features described above are indicative of a male skull. The absence of degenerative changes on the only skull joint and the assessed younger dentition, as well as the absence of obliterative changes on the cranial sutures indicate the skull of an adult younger male.

The only preserved mandible was placed in the display cabinet beside this skull. It cannot be stated with certainty that the mandible is a constituent part of the skull, particularly since it is more gracile and most probably of female gender.

THE SKULL BELONGED TO AN ADULT MALE, AGED AROUND 30 YEARS

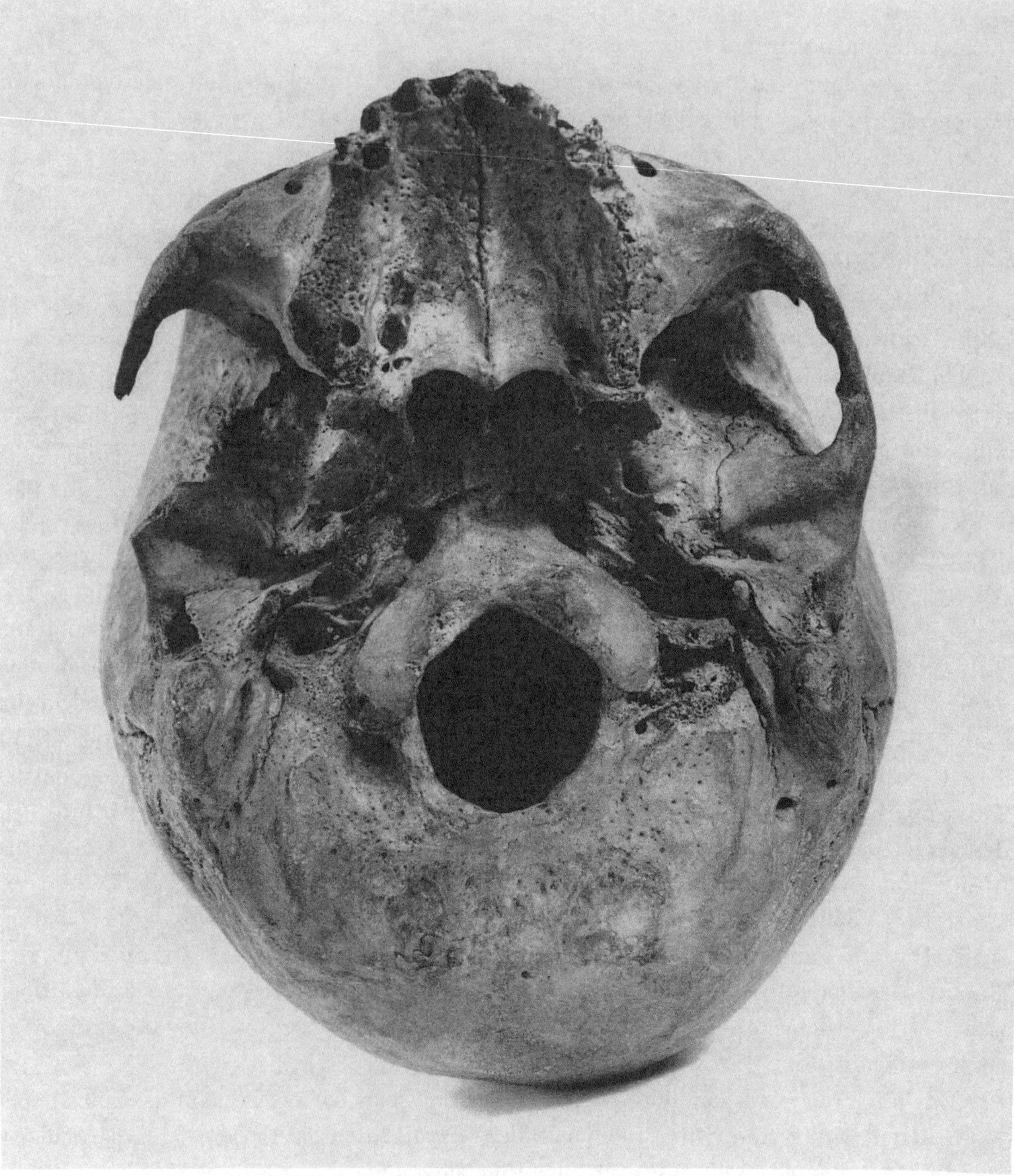

Picture 45: *Skull 9 from below. The alveolar ridge reveals the numerous atrophies of the dental alveoli. The person had lost a considerable number of teeth during lifetime*

4.2.1.9.
Skull 9 (pict. 45)

4.2.1.9.1. Determining gender by skull

Anthropological characters for determining gender

superciliary arches	moderately developed(+1)
glabella	pronounced (+l)
frontal tubers	moderately pronounced (+1)
inclination of the forehead	moderately inclined (+1)
supraorbital margin	rounded (+l)
orbital shape	rectangular (+1)
occipital relief	pronounced (+1)
ext. occipital protuberance	moderately expressed (+l)
mastoid processes	moderately expressed (+1)
cheekbones	r. broken off, l. moderately expressed
maxillary palatine process	between ellipsoid and semicircular shape
maxillary alveolar process	semicircular
thickness of calvaria	occipitally greater than frontally
facial part of skull	pronounced

Determining gender by anthropometric methods and indices

bizygomatic breadth	after reconstruction 135 mm
great occipital foramen	length 35 mm, breadth 32 mm
mastoids	both preserved, breadth r. 40 mm, l. 24 mm
breadth of anterior nasal aperture	24 mm
cranial length	184 mm
cranial breadth	146 mm
cranial base	95 mm
cranial index	79.35 (medium length skull)
orbital height	r. 32 mm, l. 31 mm
orbital breadth	r. 40 mm, l. 40 mm
orbital index	l. 77.50, r. 80.0 (male feature)

Determining gender by dental methods

Dental index after Pont:

Breadth between first premolars	39 mm. P = 48.75
Breadth between first molars	44 mm. M = 73.33

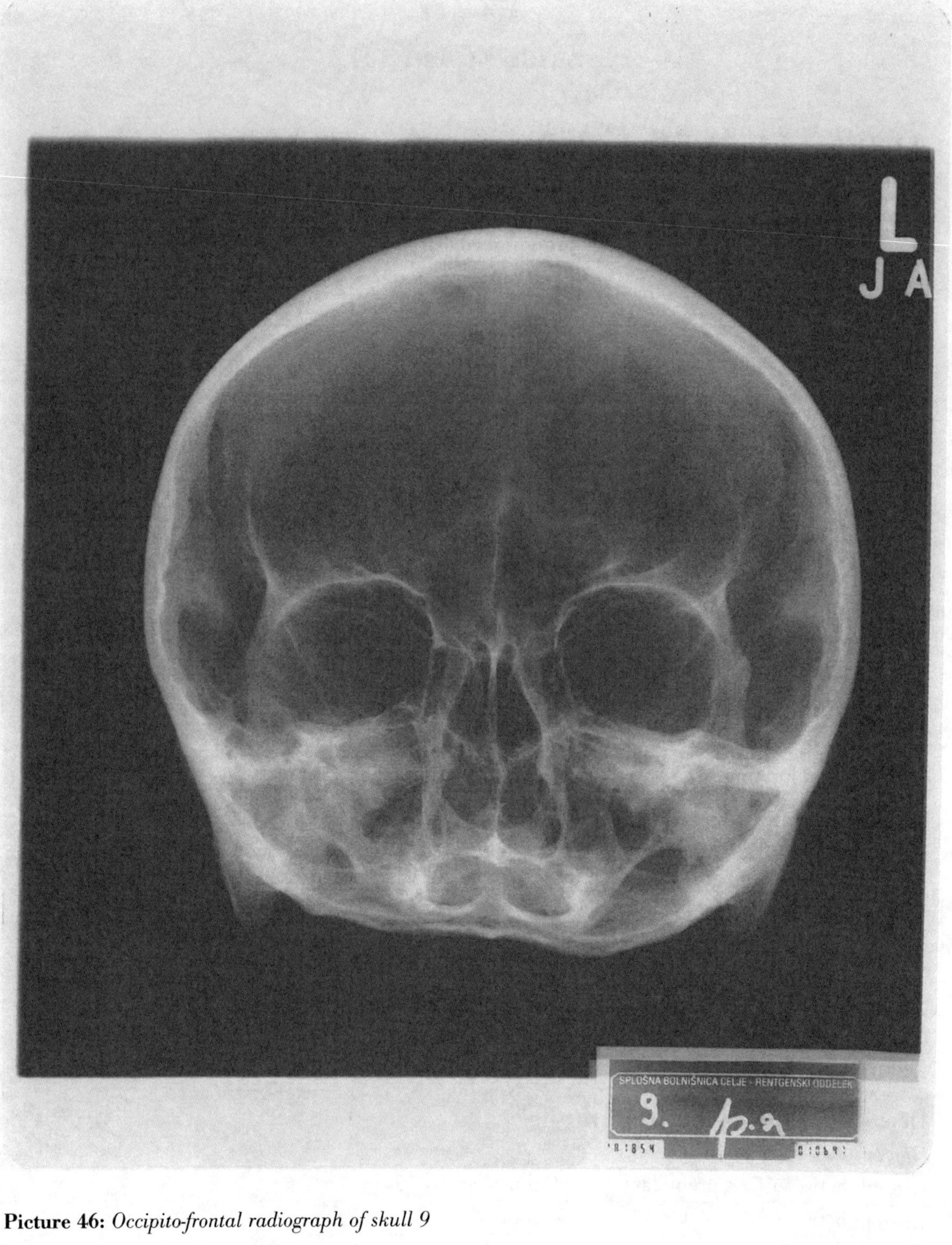

Picture 46: *Occipito-frontal radiograph of skull 9*

4.2.1.9.2. Determining age by skull

Determining age by endo- and ectocranial obliteration of the sutures

In this instance, the cranial sutures are far more pronounced than otherwise. The sagittal, coronal and lambdoidal sutures are highly visible: the first two are obliterated, while the asteric part of the third is still open. The obliterated spheno-occipital suture is also patently visible. The medial palatine suture is obliterated.

Determining age by forensic-dental methods

The maxilla does not contain any preserved teeth, and there are also few dental alveoli. The incisive foramen is preserved; the medial palatine suture is still unclosed.

To the right are preserved the alveoli of the first and second incisors, the canine, the first premolar – which has two roots – and the third molar. Between the alveoli of the first premolar and the third molar, the alveolar ridge of the maxilla is completely atrophied. On the left are preserved the alveoli of the first incisor, the canine and the first premolar. The person must have already lost the second left incisor early during his lifetime; the remaining part of the alveolar arch indicates complete sclerotization.

Maxillary alveolar process from right towards left:

	M3	M2	M1	PM2	PM1	C	I2	I1	I1	I2	C	PM1	PM2	M1	M2	M3
Maxilla:	a	a	a	a	+	+	+	+	a	a	+	+	a	a	a	a

Legend: *a = atrophic*

Determining age by X-ray methods (pict. 46)

The fronto-occipital radiograph shows a skull with thick frontal bone on the vertex. The orbits are relatively rectangular; the nasal bone is relatively broad, with the septum deviating slightly to the left. On the fronto-occipital and occipito-frontal radiographs, Pacchionian bodies are prominently visible.

On the radiograph of the base, oval and rounded shadows can be seen. To the right, by the great occipital foramen, there is a noticeable shadow with an uneven non-sclerotic margin. On account of suspicion of a solitary metastasis, a CT scan was performed. The result showed no pathological changes.

no. 9: ♂, age 30–40 years

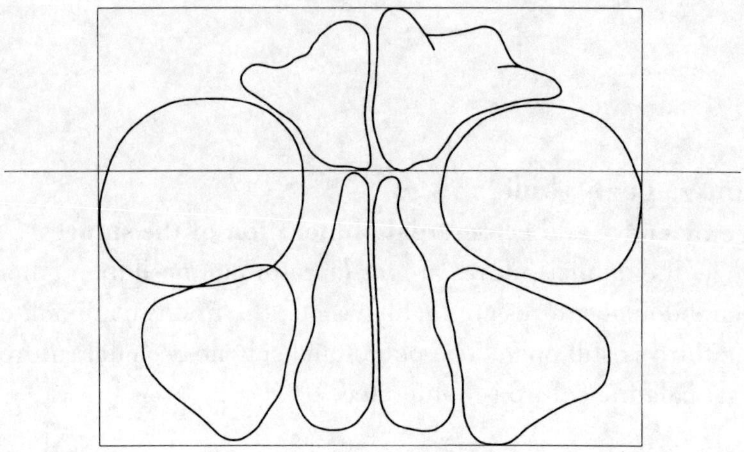

Picture 47: *Outline of the facial cavities of skull 9 from the occipito-frontal radiograph*

4.2.1.9.3. X-ray determination of the facial cavities (pict. 47)

The occipito-frontal radiograph shows that:

- the frontal sinuses are relatively large and asymmetric. The left is larger than the right; they are fan shaped;
- the orbits are rectangular in shape. After Nemeskéri they are scored as (+l);
- the maxillary sinuses are fairly symmetrical and trapezoid shape;
- the nasal cavity is symmetrical. It is scallop shaped.

Planimetric values of the facial cavities

Skull		Frontal sinuses		Orbits		Maxillary sinuses		Nasal cavity
		Right	Left	Right	Left	Right	Left	
No. 9	S	6.59	9.08	17.00	16.38	8.90	8.73	13.80
	H	3.65	4.54	4.90	4.81	4.72	4.99	5.34
	B	2.94	3.30	4.10	4.10	2.14	2.23	3.48
	T	0.72	0.63	0.72	0.81	0.45	0.36	0.63
	Σ	15.67						
	S	0.54	0.45	0.81	0.72	0.63	0.54	0.63

Legend: *S = surface, H = maximal height, B = maximal breadth, T = B/H, Σ = total surface of frontal sinuses, Sh = shape of sinuses*

4.2.1.9.4. Paleopathological methods

The left mandibular fossa is slightly porous. On the cranial occiput there are rare cancerous pathological processes. Therefore the raised suspicion of a solitary metastasis at that position was checked with a CT scan. This revealed that the change was not pathological, but rather that physiological structures were involved, known as foramina parietalia, with symmetrical, smooth and sclerotized margins.

No other paleopathological particularities were noted on skull no. 9.

4.2.1.9.5. Commentary

According to anthropometric measurements and the cranial index, the skull is of medium length. It is round shaped, compact and almost undamaged. It has most of the features of a male skull, and is between robust and gracile. Moderately expressed on the front part of the skull are: the glabella, supraorbital arches and frontal tuber. The frontal bone is thickened, and the supraorbital arches have rounded margins. The inclination of the forehead is slight. The occipital relief is slightly undulating, and the attachments of the occipital muscles are in correspondence with this. The great occipital foramen is roundish in shape; the mastoids are moderately large. On the facial part the orbits are rectangular and the orbital indices are characteristic of a male. The nasal bone is wide at the root, the nasal septum straight, and the nasal cavity triangular shape. Campers' angle indicates an orthognathic occlusion and regular positioning of the teeth. The dental index after Pont indicates a semicircular dental arch which, together with broad cheekbones, contributes to forming a wider and harmoniously structured face. During his lifetime, the person had lost a great number of teeth. For pronounced atrophy of the alveoli at least 10–20 years are needed. Was scurvy to blame, or was it the extremely poor dentition?

The occipital condyles are asymmetric: the left is lower, and hence the skull is not evenly positioned, but is inclined to the left. The man probably carried his head inclined to one side (the left occipital condyle is 7 mm lower than the right). This sign was sufficiently characteristic that it would have been mentioned in one of the descriptions, if it had been preserved. However, in the vast amount of material studied, no trace of this was found.

The left mandibular fossa is arthrotic, while occipitally the radiograph indicated a pathological process. A solitary metastasis was suspected, however this was dismissed after computerised tomography. The above record describes an adult male skull, aged approx. 40 years. On the basis of the research, the possible cause of death could not be discovered.

THE SKULL BELONGED TO AN ADULT MALE, AGED 30–40 YEARS

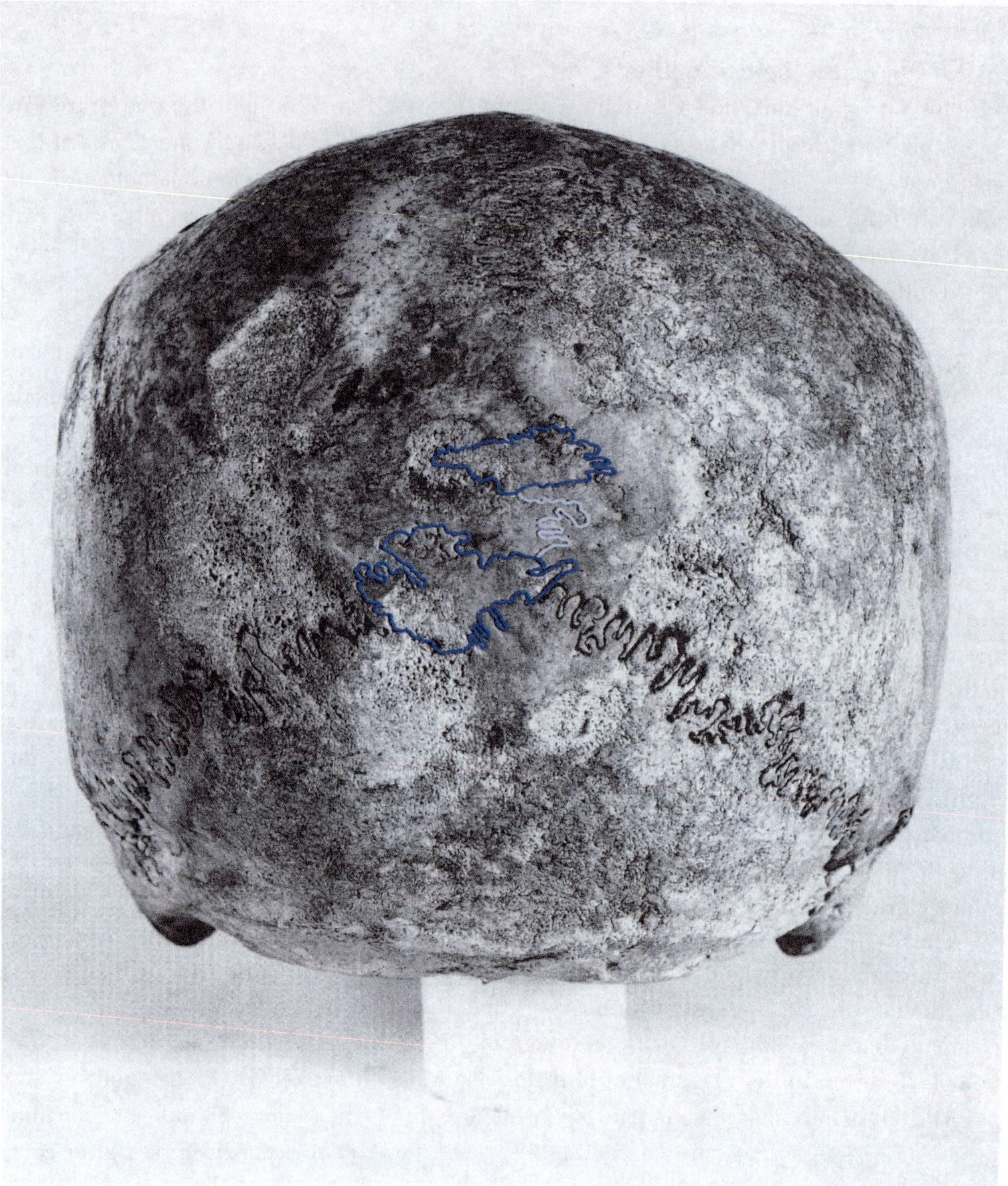

Picture 48: *Skull 10 from behind. Beside the lambda are two larger lambdoidal ossicles*

4.2.1.10.

Skull 10 (pict. 48)

4.2.1.10.1. Determining gender by skull

Anthropological characters for determining gender

superciliary arches	not developed (–2)
glabella	not expressed (–2)
frontal tubers	expressed (–1)
inclination of the forehead	almost vertical (–1)
supraorbital margin	thin, narrow and sharp (–2)
orbital shape	between rectangular and round (–1)
occipital relief	slightly expressed (–2)
ext. occipital protuberance	unexpressed (–2)
mastoid processes	r. larger than l. (–1)
cheekbones	gracile (–2)
maxillary palatine process	ellipsoid
maxillary alveolar process	elliptic
thickness of calvaria	uniform
facial part of skull	relatively small

Determining gender by anthropometric methods and indices

bizygomatic breadth	127 mm
great occipital foramen	length 33 mm, breadth 28 mm
mastoids	breadth r. 30 mm, l. 24 mm
breadth of anterior nasal aperture	27 mm
cranial length	169 mm
cranial breadth	140 mm
cranial index	82.84 (short skull)
cranial base	89 mm
orbital height	r. 37 mm, l. 37 mm
orbital breadth	r. 40 mm, l. 40 mm
orbital index	r. and l. 92.50 (female feature)

Determining gender by dental methods

Dental index after Pont:

Breadth between first premolars	32 mm. P = 40
Breadth between first molars	38 mm. M = 63.33

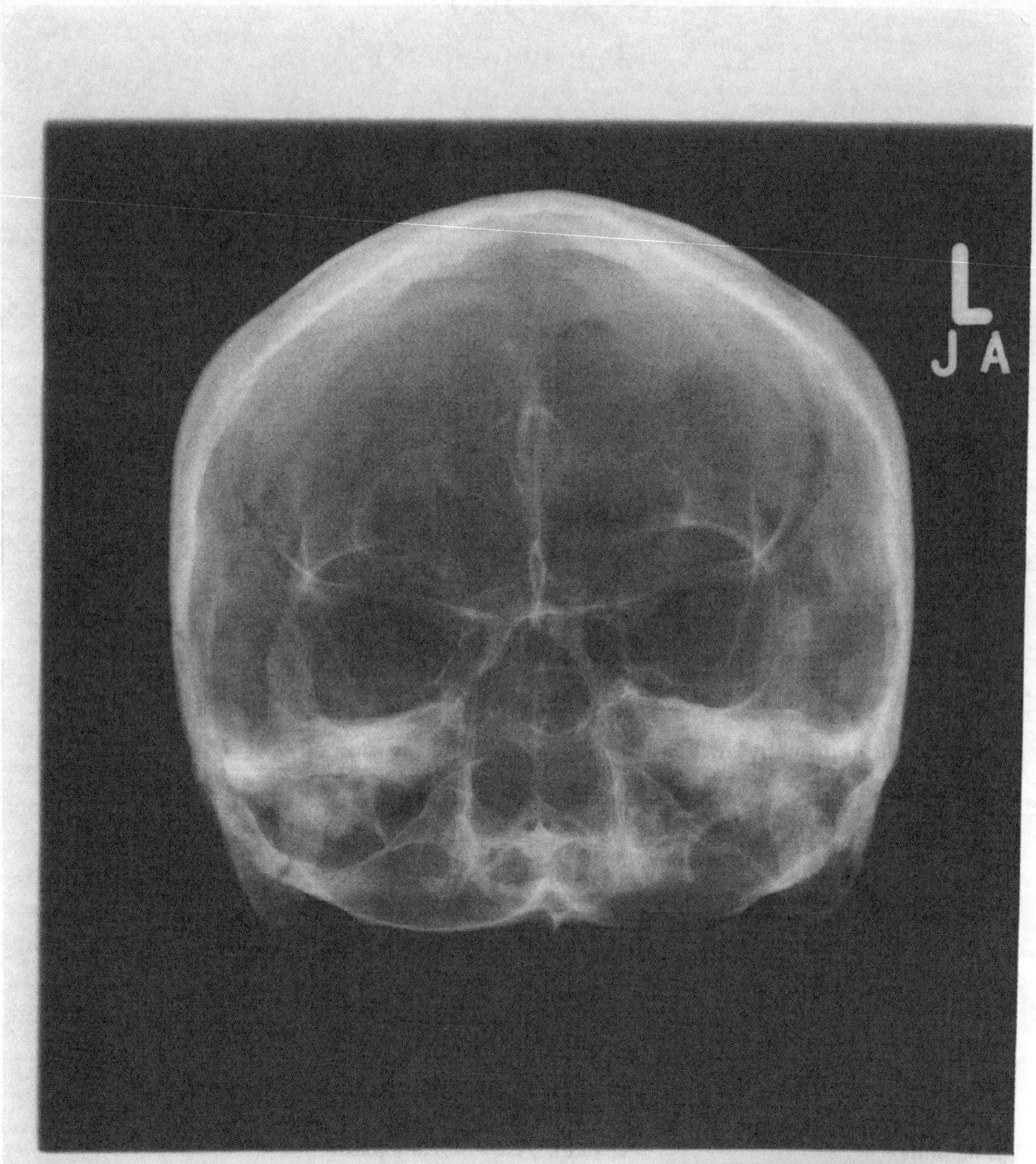

Picture 49: *Occipito-frontal radiograph of skull 10*

4.2.1.10.2. Determining age by skull

Determining age by endo- and ectocranial obliteration of the sutures

All the cranial sutures still show characters of obliteration: both sagittal and coronal, and particularly strongly on the lambdoidal. The temporo-occipital suture and both sutures on the hard palate are in similar state.

Determining age by forensic-dental methods

The maxillary alveolar ridge is protrudent. The teeth had fallen out after death. The incisive foramen is well exposed. The hard palate is ellipsoid and narrow; the palatine sutures – medial and transversal – are not obliterated. The palatine torus is also highly pronounced.

To the right, on the alveolar ridge, the dental alveoli from the first incisor to the first molar are preserved; the remaining part of the right maxillary tuber is broken off.

To the left, the dental alveoli are preserved from the first incisor to the second molar.

Maxillary alveolar process from right towards left:

	M3	M2	M1	PM2	PM1	C	I2	I1	I1	I2	C	PM1	PM2	M1	M2	M3
Maxilla:	b	b	+	+	+	+	+	+	+	+	+	+	+	+	+	a

Legend: *a = atrophic, b = broken off*

Determining age by X-ray methods (pict. 49)

The fronto-occipital radiograph shows a skull with orbits shaped between rectangular and rounded. The defects of the frontal bone are projected into the sinuses. The nasal bone is moderately broad, with the nasal septum preserved only in the top part. The maxilla is strongly protrudent. On the left and right side images a partial aperture may be seen, more of the coronal and less of the lambdoidal suture, and also a somewhat larger left mastoid. A more precise examination of the septum in the left mastoid as compared with the right indicated increased sclerotization, which may be attributed to subacute or chronic mastoiditis of the adult age.

No other particular features were obtained from the radiographs.

no. 10: ♀, 30–40 years

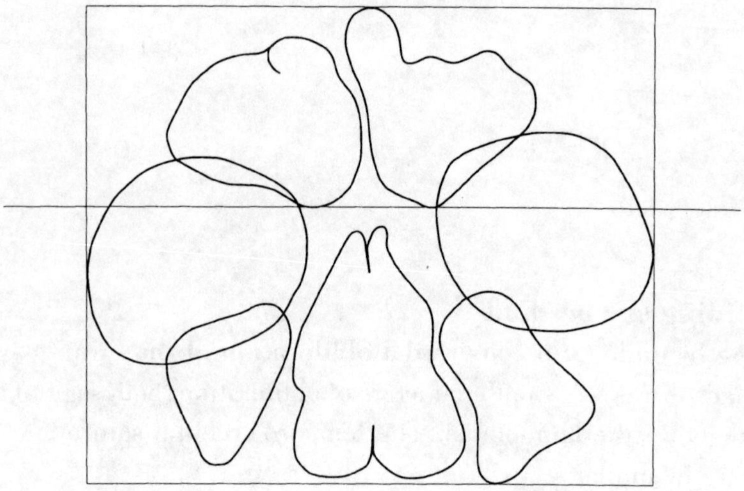

Picture 50: *Outline of the facial cavities of skull 10 from the occipito-frontal radiographs*

4.2.1.10.3 X-ray determination of the facial cavities (pict. 50)

The occipito-frontal radiograph shows that:

- the frontal sinuses are relatively large, asymmetrical and fan shaped;
- the orbits are asymmetrical, the left is larger than the right, and in shape the right is more rounded, the left more rectangular. After Nemeskéri, they are scored as (–1);
- the maxillary sinuses are asymmetrical. They resemble a trapezoid shape;
- the nasal cavity is symmetrical, and is scallop shaped with a serrated margin.

Planimetric values of the facial cavities

Skull		Frontal sinuses		Orbits		Maxillary sinuses		Nasal cavity
		Right	Left	Right	Left	Right	Left	
No. 10	S	8.65	8.29	12.84	14.20	11.84	9.65	12.29
	H	3.82	4.00	4.10	4.28	4.73	4.37	5.10
	B	3.10	3.37	3.83	3.92	3.55	2.92	3.28
	T	0.73	0.73	0.82	0.82	0.64	0.64	0.55
	Σ	6.93						
	Sh	0.73	0.46	0.73	0.82	0.64	0.73	0.55

Legend: *S = surface, H = maximal height, B = maximal breadth, T = B/H, Σ = total surface of frontal sinuses, Sh = shape of sinuses*

4.2.1.10.4. Paleopathological methods

Among the pathological changes on skull no. 10, we can include only one, most probably inflammatory, process of subchronic or chronic nature, which has developed along the left mastoid. As a consequence, the person may have had poor hearing in the left ear.

4.2.1.10.5. Commentary

According to the anthropometric measurements and the cranial index, this skull is short. It is oval shaped, gracile and partly damaged. By features of shape it is a typical female skull with unpronounced glabella, supraorbital arches and frontal tubers. The forehead is gently inclined. The frontal bone is damaged and on both sides one can see into the interior of the frontal sinuses. The unpronounced occiptal protuberance, the sparse occipital relief with unexpressed muscle attachments and the more rounded foramen occipitale are also feminine traits. Both mastoids are preserved; the left is slightly larger than the right and displays signs of chronic inflammation, on account of which the person most probably had poor hearing in the left ear.

The shape of the orbits is between round and rectangular, and by the orbital index they belong to a female. The nasal root is broad; the nasal bone and septum, however, are broken off. The dental index after Pont indicates a narrower dental arch. Campers' angle expresses considerable prognathism of the maxilla. Since the cheekbones are also narrow, the traits described indicate a narrower face with a protuberant maxilla and with it also protruding, otherwise intravitally healthy teeth.

The fact that many of the cranial sutures are still open was a help in determining the age. The mandibular fossae also do not display degenerative changes. The teeth fell out mainly after death. These features speak in favour of a younger person. Among the illness induced changes – apart from the suspected hearing difficulties – no particular features were noted.

The above record undoubtedly indicates the skull of a younger adult woman. The cause of death cannot be determined from the skull.

THE SKULL BELONGED TO AN ADULT FEMALE, AGED 30–40 YEARS

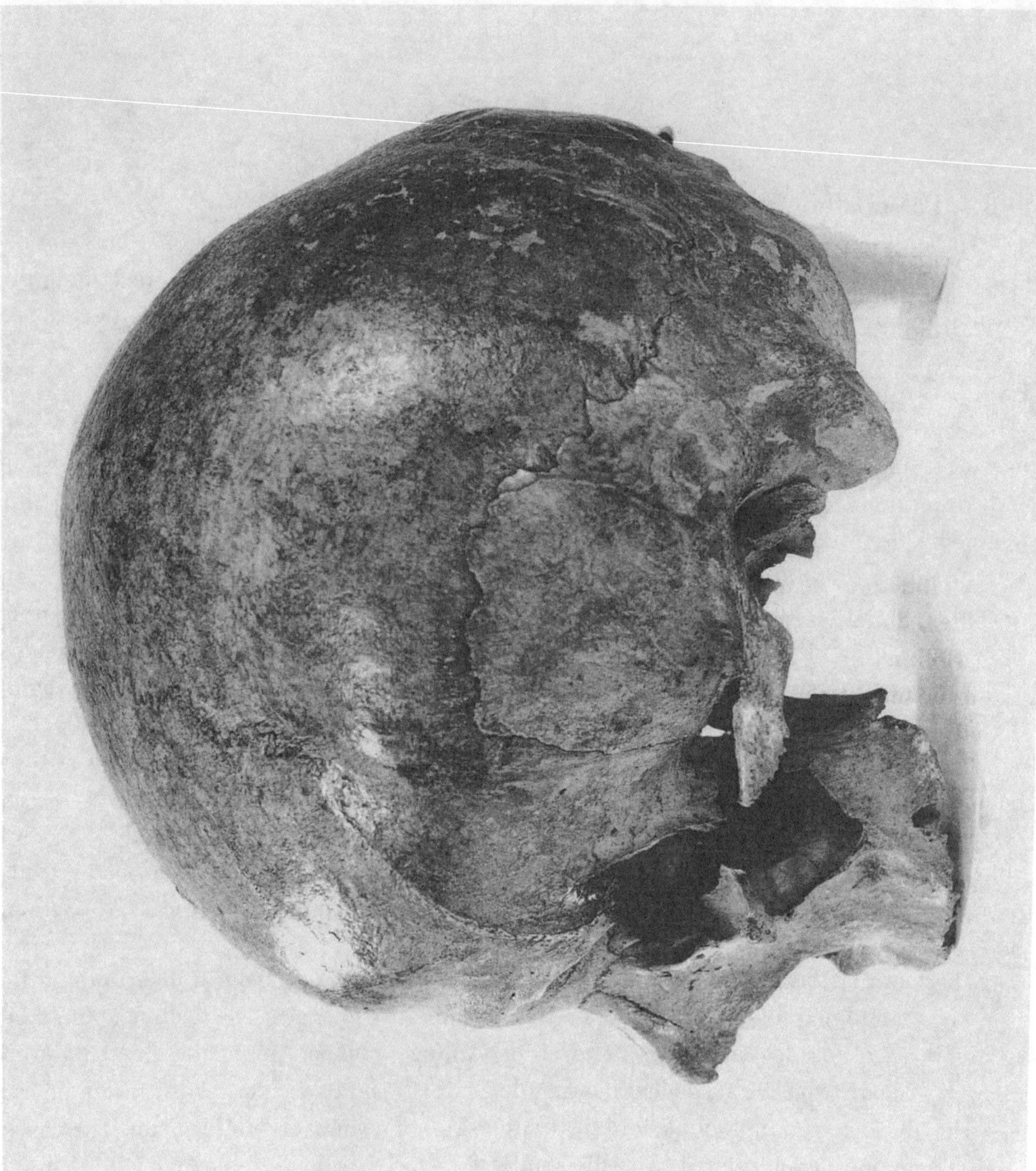

Picture 51: *Skull 11, side view. It has a strong nasal bone, most probably hook shape*

4.2.1.11
Skull 11 (pict. 51)

4.2.1.11.1. Determining gender by skull

Anthropological characters for determining gender

superciliary arches	unpronounced on both sides (0)
glabella	unpronounced on both sides (0)
frontal tubers	expressed (–1)
inclination of the forehead	almost vertical (–1)
supraorbital margin	unpronounced (–1)
orbital shape	more rounded (–1)
occipital relief	bone broken off
ext. occipital protuberance	bone broken off
mastoid processes	missing
cheekbones	broken off
maxillary palatine process	ellipsoid
maxillary alveolar process	semicircular
thickness of calvaria	frontally 6 mm, occipitally 5 mm
facial part of skull	relatively strong

Determining gender by anthropometric methods and indices

bizygomatic breadth	damaged
great occipital foramen	missing
mastoids	only the left preserved, breadth 28 mm
breadth of anterior nasal aperture	25 mm
cranial length	after reconstruction approx. 175 mm
cranial breadth	after reconstruction 147 mm
cranial base	undeterminable due to damage
Campers' angle	approx. 80° (harmoniously structured face)
cranial index	84 (short skull)
orbital height	r. 38 mm, l. damaged
orbital breadth	r. 44 mm, l. damaged
orbital index	r. 86.36, l. undeterminable due to damage.
	Value characteristic of a female skull.

Determining gender by dental methods

Dental index after Pont:

Breadth between first premolars	33 mm. P = 41.25
Breadth between first molars	41 mm. M = 68.33

The palate is therefore ellipsoid shape.

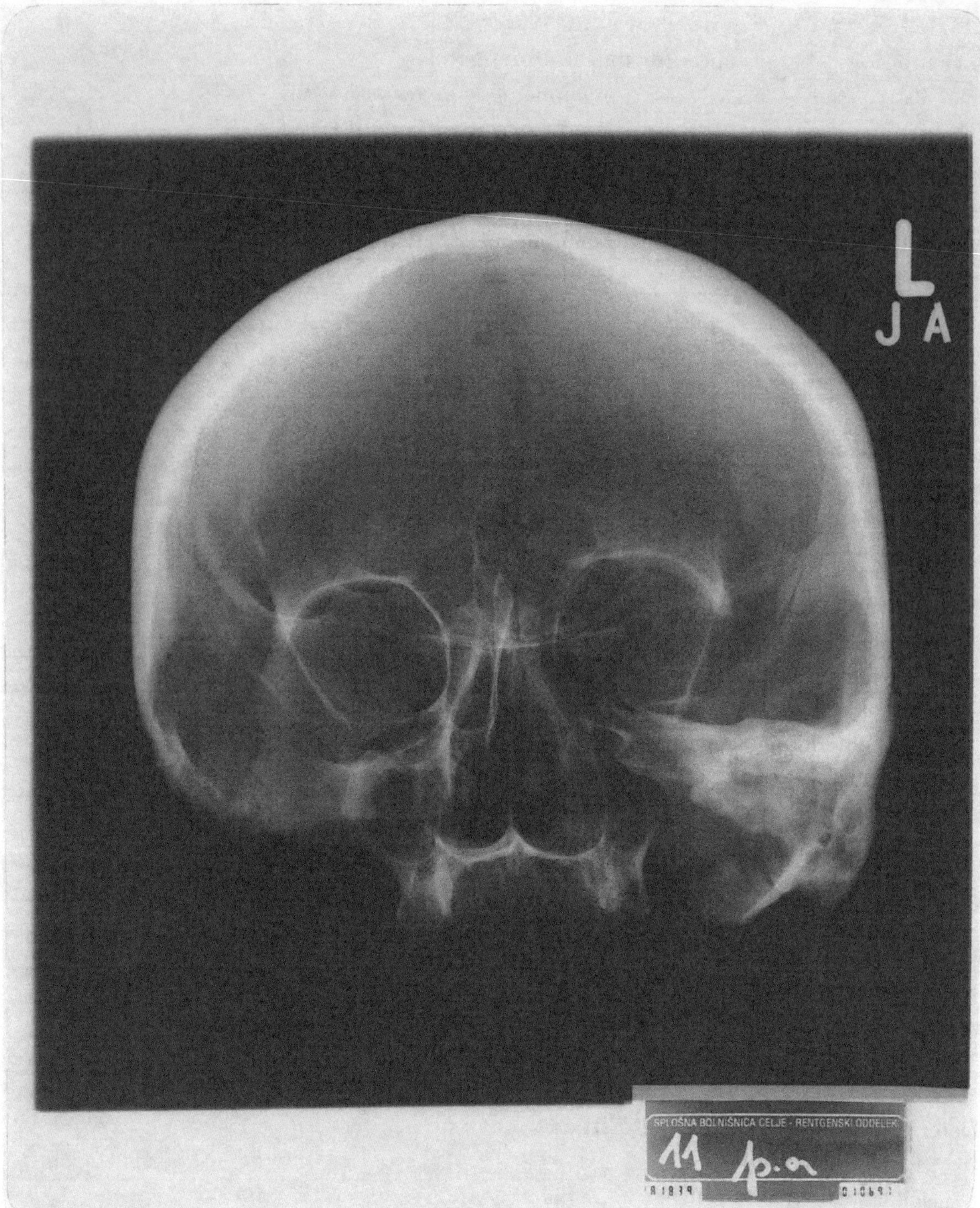

Picture 52: *Occipito-frontal radiograph of skull 11*

4.2.1.11.2. Determining age by skull

Determining age by endo- and ectocranial obliteration of the sutures

The sagittal suture is completely obliterated, the coronal in the central part, the lambdoidal likewise only in the central part. The facial sutures are also completely obliterated.

The features of the sutures described indicate an elderly person, certainly older than 50 years.

Determining age by forensic-dental methods

The maxilla is without teeth, however the alveoli of all 16 teeth are preserved. The person clearly had very healthy teeth, since most of them had fallen out only after death. The alveolus of the first left molar opens into the maxillary sinus. The palate is ellipsoid shape. The medial palatine suture is still not closed; by its throat part the palatine torus is slightly developed. The incisive foramen is medium pronounced.

Maxillary alveolar process

On the right the dental alveoli from 1 to 8 are well preserved, with the exception of the second premolar.

On the left the dental alveoli are well preserved up to the second premolar; the remaining part of the alveolar process is partially atrophied. The alveolus of the first molar is completely atrophic. The molars, therefore, were most probably lost during this person's lifetime.

Maxillary alveolar process from right towards left:

	M3	M2	M1	PM2	PM1	C	I2	I1	I1	I2	C	PM1	PM2	M1	M2	M3
Maxilla:	+	·+	+	a	+	+	+	+	+	+	+	+	+	a	pa	pa

Legend: *a = atrophic, pa = partially atrophic*

Determining age by X-ray methods (pict. 52)

The radiographs reveal large bone defects. On the fronto-occipital view one can see smaller, more rounded orbits, a highly prominent nasal bone with a proximally preserved nasal septum, which is distally broken off and deviating to the right. Frontally, there are some visible smaller Pacchionian bodies.

At the front and back the thickness of the bone is approximately equal; the sulci of the diploic veins are pronounced. The sagittal and coronal sutures are obliterated. The lambdoidal suture is still partially open. There are no other particular X-ray features.

no. 11: ♀, aged around 40

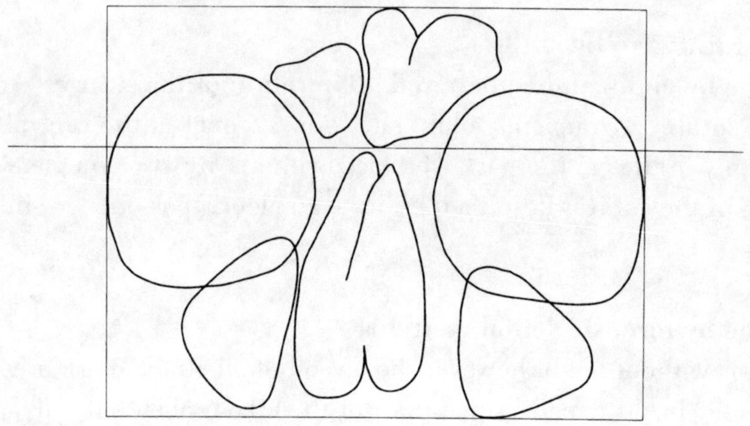

Picture 53: *Outline of the facial cavities of skull 11 from the occipito-frontal radiograph*

4.2.1.11.3. X-ray determination of the facial cavities (pict. 53)

The occipito-frontal radiograph shows that:

- the frontal sinuses are asymmetric, the left is larger than the right. The left is more reminiscent of a leaf shape, the right of a fan shape;
- the orbits are more rounded in shape. After Nemeskéri they are scored as (–1);
- the maxillary sinuses are fairly symmetrical. In shape they resemble a triangle;
- the nasal cavity is slightly asymmetric and pear shaped.

Planimetric values of the facial cavities

Skull		Frontal sinuses		Orbits		Maxillary sinuses		Nasal cavity
		Right	Left	Right	Left	Right	Left	
No. 11	S	2.14	6.23	18.43	17.27	7.12	6.59	11.04
	H	1.96	3.21	4.90	4.72	3.83	4.19	5.34
	B	1.43	2.67	4.45	4.19	2.23	1.96	2.94
	T	0.63	0.72	0.81	0.81	0.54	0.45	0.54
	Σ	8.37						
	Sh	0.72	0.54	0.89	0.89	0.72	0.63	0.54

Legend: *S = surface, H = maximal height, B = maximal breadth, T = B/H, Σ = total surface of frontal sinuses, Sh = shape of sinuses*

4.2.1.11.4. Paleopathological methods

The left mandibular fossa reveals characters of arthropathy. There are no other paleopathological changes.

4.2.1.11.5. Commentary

The skull was posthumously considerably damaged; it lacks the entire base, the right temporal bone and occiput, and therefore certain important features cannot be assessed. According to the cranial index it is short, and rounded in shape. The glabella, supraorbital arches, the frontal and parietal tubers and the inclination of the forehead are indicative of feminine features. The orbital margin is also unpronounced. On account of the damaged state of the skull, it is not possible to assess the important features which are provided by the occipital protuberance, occipital relief and foramen.

The orbits are round and smaller, the right orbital index indicates a female type of orbit, but the left, on account of the damage, cannot be determined. The nasal bone is strong – the nose was probably hook shaped – the nasal septum distally broken off and deviating to the right. The right cheekbone is missing, while on the left there is no arch. The cheekbones also do not reveal male features. Campers' angle indicates an orthognathy; the dental index after Pont reflects a semicircular dental arch. The characters described on the face are indicative of a broader, harmoniously structured face with healthy teeth which, with the exception of two molars, fell out posthumously. Both upper wisdom teeth had also grown out. Only the left mastoid, which is small, is preserved. There is a very strong temporal line, which represents the measure for the prominence of the masticatory muscles. The left mandibular fossa is arthrotic, the right smooth and without special features. The calvaria, both frontally and occipitally, is of approximately equal thickness, which is indicative of a younger female.

The described features of skull 11 are associated with the female gender. Of considerable assistance in determining the age were data on the degenerative changes on the left maxillary joint, the very healthy dentition – with the exception of the two early lost molars – and also the growing out of the two wisdom teeth, which usually appear between the ages of 16 and 30 years. Among the cranial sutures, the lambdoidal is still partly open, while the facial sutures are obliterated. The above description conforms to that of an older adult woman.

It appears that there was earth or water in the skull.

THE SKULL BELONGED TO AN ADULT WOMAN, AGED ABOUT 40 YEARS

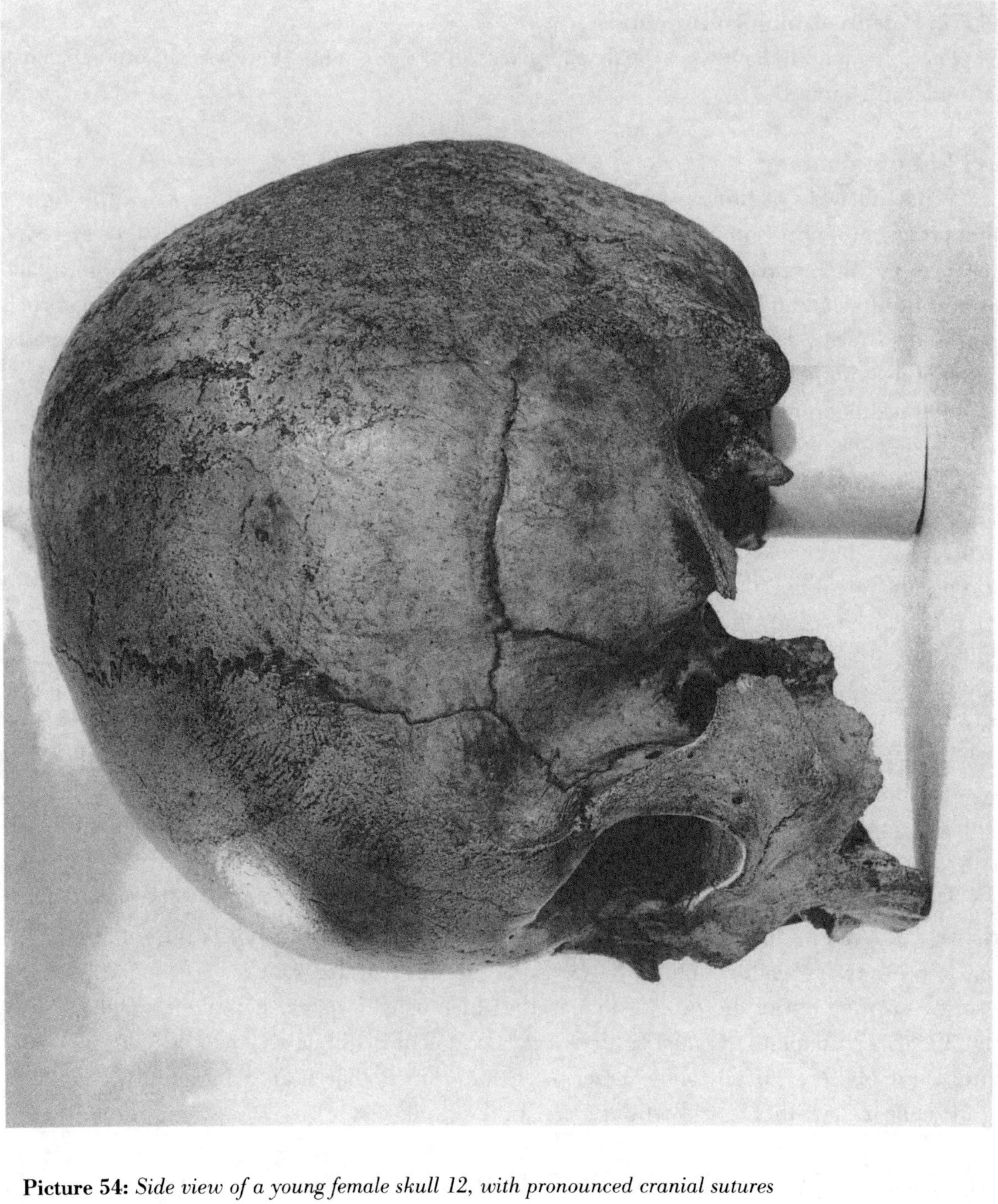

Picture 54: *Side view of a young female skull 12, with pronounced cranial sutures*

4.2.1.12.
Skull 12 (pict. 54)

4.2.1.12.1. Determining skull by gender

Anthropological characters for determining gender

superciliary arches	expressed (–2)
glabella	undeveloped (–2)
frontal tubers	expressed (–2)
inclination of the forehead	almost vertical (–2)
supraorbital margin	thin and sharp
orbital shape	rounded (–1)
occipital relief	scanty (–2)
ext. occipital protuberance	slightly expressed (–2)
mastoid processes	slightly expressed (–2)
cheekbones	l. broken off, r. gracile (–2)
maxillary alveolar process	semicircular
maxillary palatine process	elliptic
thickness of calvaria	uniform
facial part of skull	relatively small

Determining gender by anthropometric methods and indices

bizygomatic breadth	after reconstruction 115 mm
great occipital foramen	length 32 mm, breadth 28 mm
mastoids	both preserved, breadth r. 30 mm, l. 16 mm
breadth of anterior nasal aperture	24 mm
cranial length	157 mm
cranial breadth	133 mm
cranial base	88 mm
Campers` angle	80 degrees
cranial index	84.71 (short skull)
orbital height	r. 31 mm, l. 31 mm
orbital breadth	r. 36 mm, l. 36 mm
orbital index	r. and l. 86.11

Determining gender by dental methods

Dental index after Pont:

Breadth between first premolars	35 mm. P = 43.75
Breadth between first molars	38 mm. N = 63.33

The maxillary palatine process is therefore ellipsoid, while the maxillary alveolar process is semicircular.

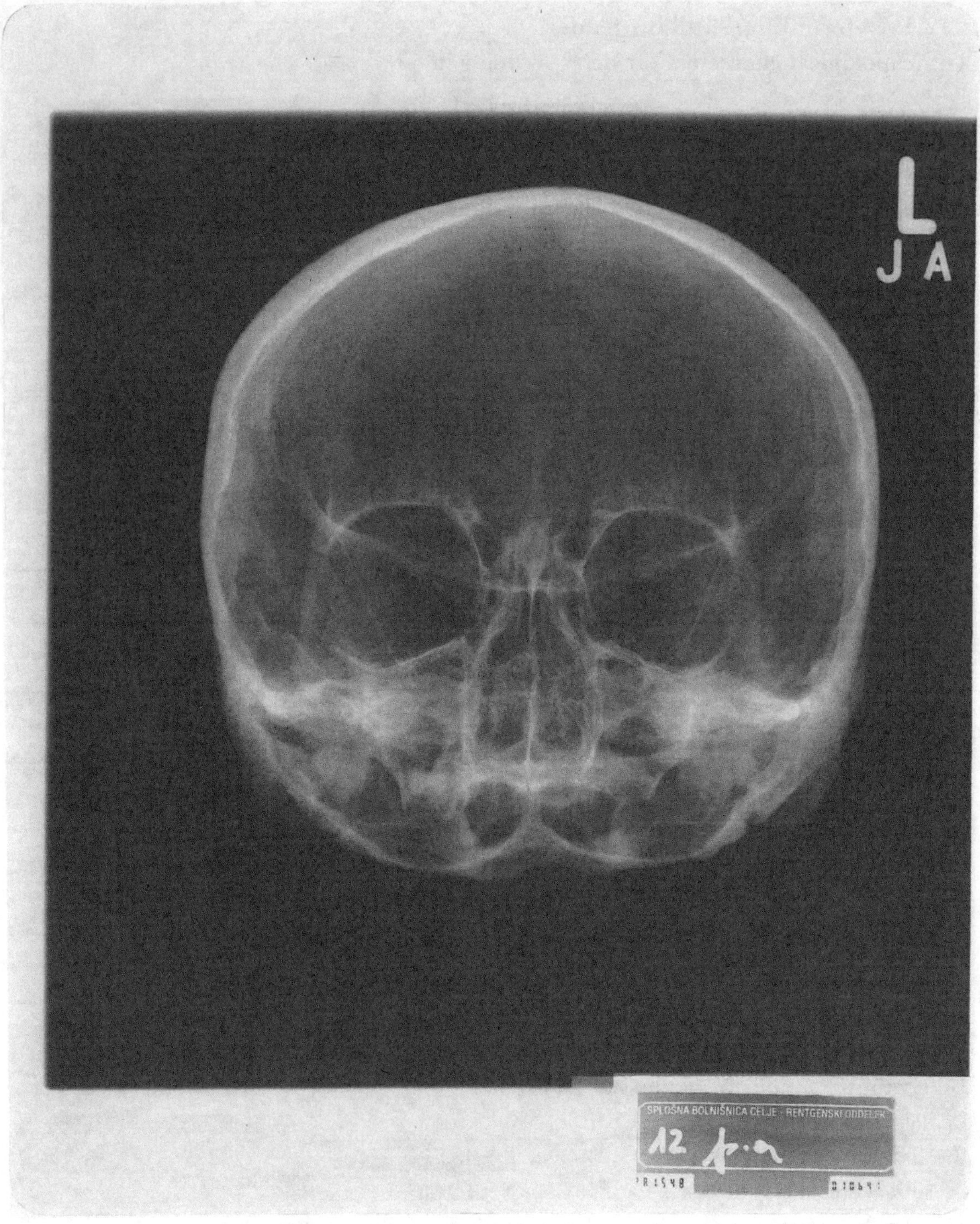

Picture 55: *Occipito-frontal radiograph of skull 12*

4.2.1.12.2. Determining age by skull

Determining age by endo- and ectocranial obliteration of the sutures

The cranial sutures are still in the process of closing, not only the sagittal but also the coronal and lambdoidal. The same holds true for the temporo-occipital suture and both sutures on the hard palate.

The assessed state of the skull sutures indicates an adult person aged about 30 years.

Determining age by forensic-dental methods

The maxilla has a toothless alveolar ridge. The alveolar process is protruding. The incisive foramen is expressed, and the hard palate complete. The medial palatine suture is open; the palatine torus highly expressed frontally, and distally becoming lost.

Maxillary alveolar process

has dental alveoli preserved from 1 to 7, both right and left; the third molar left is retained.

On the right side, the process at the place of the canine and first premolar is damaged or atrophied.

An orthopan X-ray was also performed.

Maxillary alveolar process from right towards left:

	M3	M2	M1	PM2	PM1	C	I2	I1	I1	I2	C	PM1	PM2	M1	M2	M3
Maxilla:	r	+	+	+	a	a	+	+	+	+	+	+	a	+	+	+

Legend: *a = atrophic, r = retained tooth*

Determining age by X-ray methods (pict. 55)

The fronto-occipital radiograph shows a gracile skull with round orbits; the nasal bone is gracile, the nasal septum straight and fully preserved.

On the left and right side views, the two already considerably obliterated coronal sutures may be seen. At the back it can be noted that the lambdoidal suture is also moderately obliterated. The thickness of the calvaria, both front and back, is approximately equal.

No other special features can be noted from the X-rays.

no. 12: ♀, aged around 30

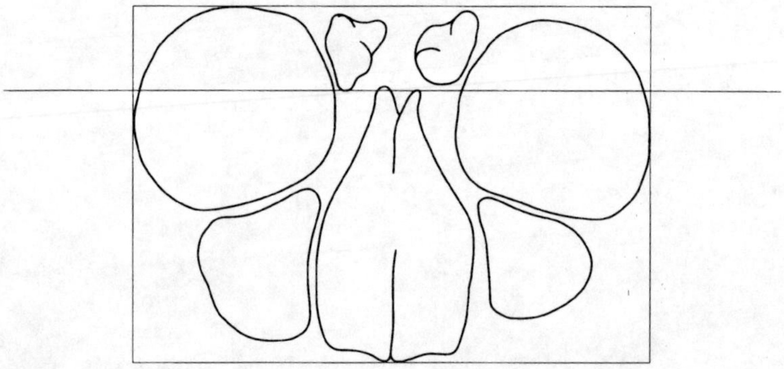

Picture 56: *Outline of the facial cavities of skull 12 from the occipito-frontal radiograph*

4.2.1.12.3. X-ray determination of the facial cavities (pict. 56)

The occipito-frontal radiograph shows that:

- the frontal sinuses are undeveloped, aplastic. Both ethmoidal sinuses, however, are apparent;
- the orbits are more rounded in shape. After Nemeskéri they are scored as (–1);
- the maxillary sinuses are relatively symmetrical and triangular in shape;
- the nasal cavity is symmetrical, resembling a pear shape.

Planimetric values of the facial cavities

Skull		Frontal sinuses		Orbits		Maxillary sinuses		Nasal cavity
		Right	Left	Right	Left	Right	Left	
No. 12	S	0.54	0.63	15.05	14.78	4.90	4.81	10.60
	H	1.07	1.25	4.45	4.37	2.76	2.85	4.90
	B	0.63	0.63	4.01	4.01	2.05	2.14	3.03
	T	0.54	0.45	0.81	0.81	0.63	0.63	0.54
	Σ	1.16						
	Sh	0.63	0.63	0.89	0.89	0.81	0.72	0.63

Legend: *S = surface, H = maximal height, B = maximal breadth, T = B/H, Σ = total surface of frontal sinuses, Sh = shape of sinuses*

4.2.1.12.4. Paleopathological methods

Macroscopically, the sieve-like changes on both sides to the orbital roofs attracted our attention. The roofs of both orbits are more or less uniformly thickened and finely perforated (pict. 57). Frontally, left and right above the orbits, there are minute perforations, similar to those which can be seen on the skull vault and occiput. On the fronto-occipital radiograph, the skull reveals strong thickening of the frontal bone and aplasia of the frontal sinuses. Frontal hyperostosis may be a physiological occurrence, however in this instance it is notable that only the ethmoidal sinuses are slightly developed in the frontal bone. No further particularities were evident from the other radiographs.

Both mandibular fossae are smooth, without surface changes or erosions.

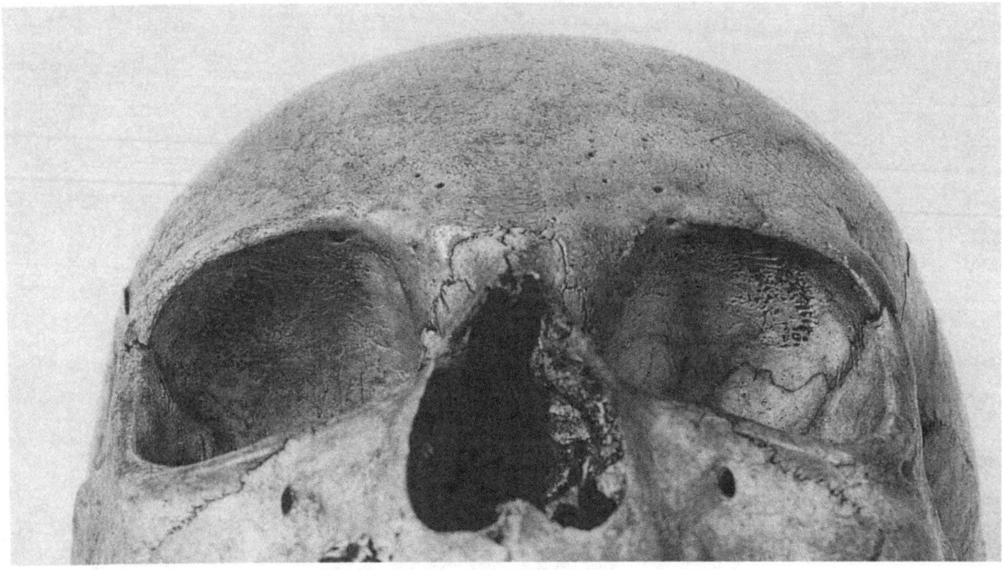

Picture 57: *Cribra orbitalia. Sieve-like orbital roofs of skull 12*

4.2.1.12.5. Commentary

This gracile skull, according to the cranial index, is short. It is very light, ovaloid and almost undamaged. The majority of the features undoubtedly indicate the female gender: glabella, supraorbital arches, frontal tuber, cheekbones and frontal inclination of the forehead. Equally convincing confirmation is given by: the occipital relief, occipital protuberances and the mastoids. The great occipital foramen is also more ellipsoid. The orbits are more rounded, and the orbital indices reflect feminine provenance. The nasal bone is partially broken off, the nasal septum is straight and fully preserved. Campers' angle indicates dysgnathia: the maxilla is slightly protuberant. The maxillary palatine process is ellipsoid, the alveolar process semicircular. Hence, the face was narrowish, the forehead relatively high, the cheekbones were unpronounced, and the upper teeth somewhat outward projecting. The dental alveoli indicate young healthy teeth which fell out posthumously. The right upper wisdom tooth has not grown out, while the alveoulus of the left one is preserved.

The above described characters represent the prototype of a gracile female skull.

In determining the age, we were aided by the following traits: the mandibular fossae of the temporomandibular joint are smooth and without degenerative changes. The dental alveoli are intact, indicating that the teeth fell out posthumously; one of the two upper wisdom teeth did not grow out. The thickness of the calvaria is greater on the front. The cranial sutures are partially obliterated. The skull has not developed frontal sinuses (aplasia); small ethmoidal sinuses are, however, visible.

It is estimated that this is the skull of a woman, aged approximately 30 at her death.

The important pathological features include the perforated roofs of both orbits, known as cribra orbitalia. According to the conjectures of the paleopathologists, this sign is an indication of anaemia and iron deficiency, and also of under-nutrition. It is conjectured that this sign had fatal consequences for the person in question.

THE SKULL BELONGED TO A YOUNGER ADULT WOMAN, AGED AROUND 30 YEARS

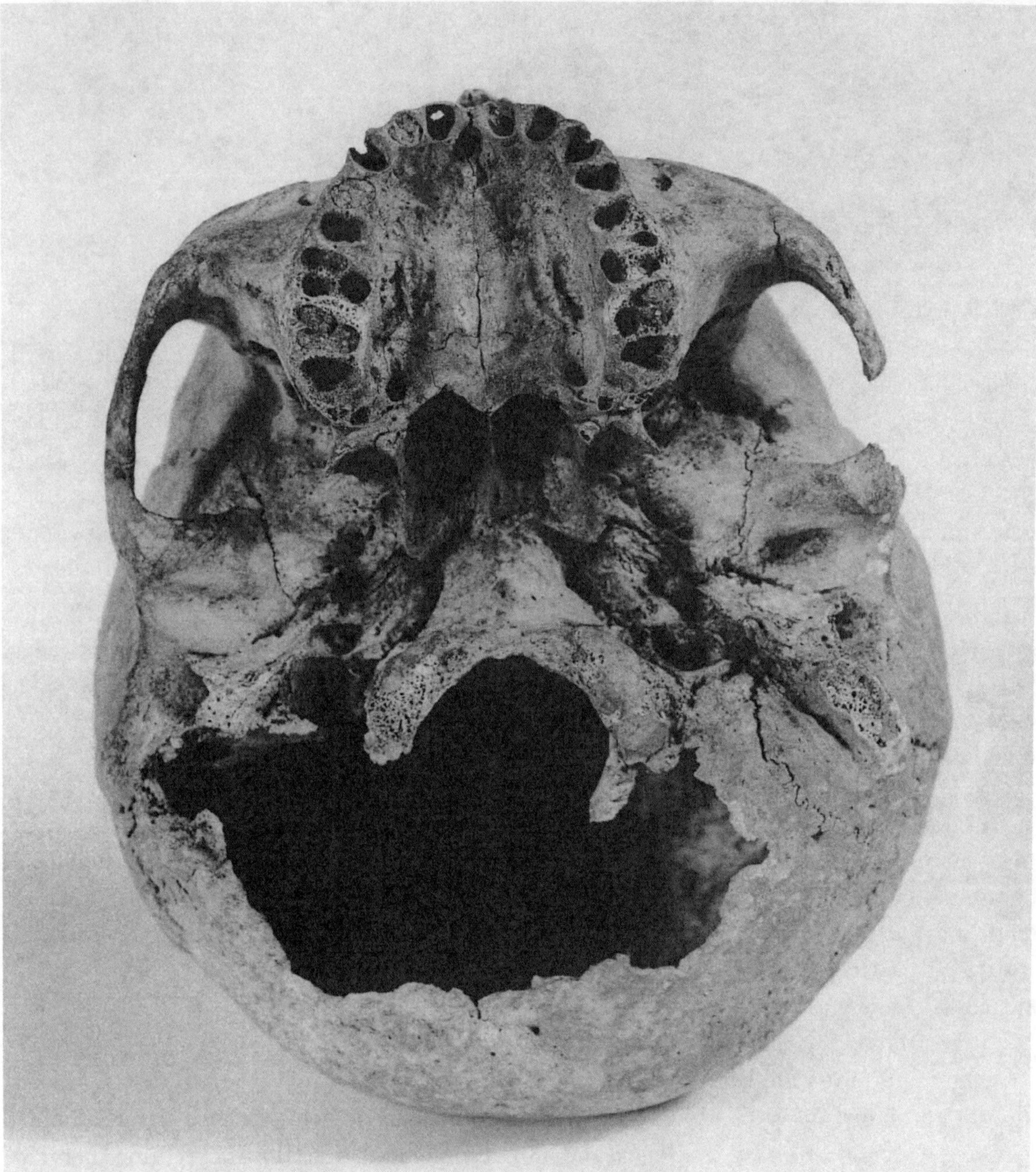

Picture 58: *Skull 13 from below reveals considerable posthumous damage to the cranial base and the semicircular dental arch with 16 dental alveoli. The teeth fell out posthumously*

4.2.1.13.

Skull 13 (pict. 58)

4.2.1.13.1. Determining gender by skull

Anthropological characters for determining gender

superciliary arches	unexpressed on both sides (0)
glabella	unexpressed on both sides (0)
frontal tubers	expressed (+1)
inclination of the forehead	markedly inclined (+1)
supraorbital margin	highly expressed (+2)
orbital shape	rectangular (+l)
occipital relief	pronounced (+1)
ext. occipital protuberance	pronounced (+l)
mastoid processes	missing
cheekbones	l. broken off, r. moderately developed (0)
maxillary palatine process	ellipsoid
maxillary alveolar process	semicircular
thickness of calvaria	frontally 4 mm, occipitally 5 mm
facial part of skull	relatively small

Determining gender by anthropometric methods and indices

bizygomatic breadth	after reconstruction 32 mm
	partially broken off, rounded,
	breadth after reconstruction 132 mm
mastoids	missing
breadth of anterior nasal aperture	28 mm
cranial length	after reconstruction 179 mm
cranial breadth	150 mm
cranial base	100 mm
Campers` angle	about 80 degrees
cranial index	83.88 (short skull)
orbital height	r. 38 mm, l. 38 mm
orbital breadth	r. 40 mm, l. 40 mm
orbital index	95

Determining gender by dental methods

Dental index after Pont:

Breadth between first premolars	40 mm. P = 50
Breadth between first molars	53 mm. P = 80

The maxillary palatine process is, therefore, large, more rounded in shape, while the maxillary alveolar process is semicircular in shape. The traits indicate a broad face.

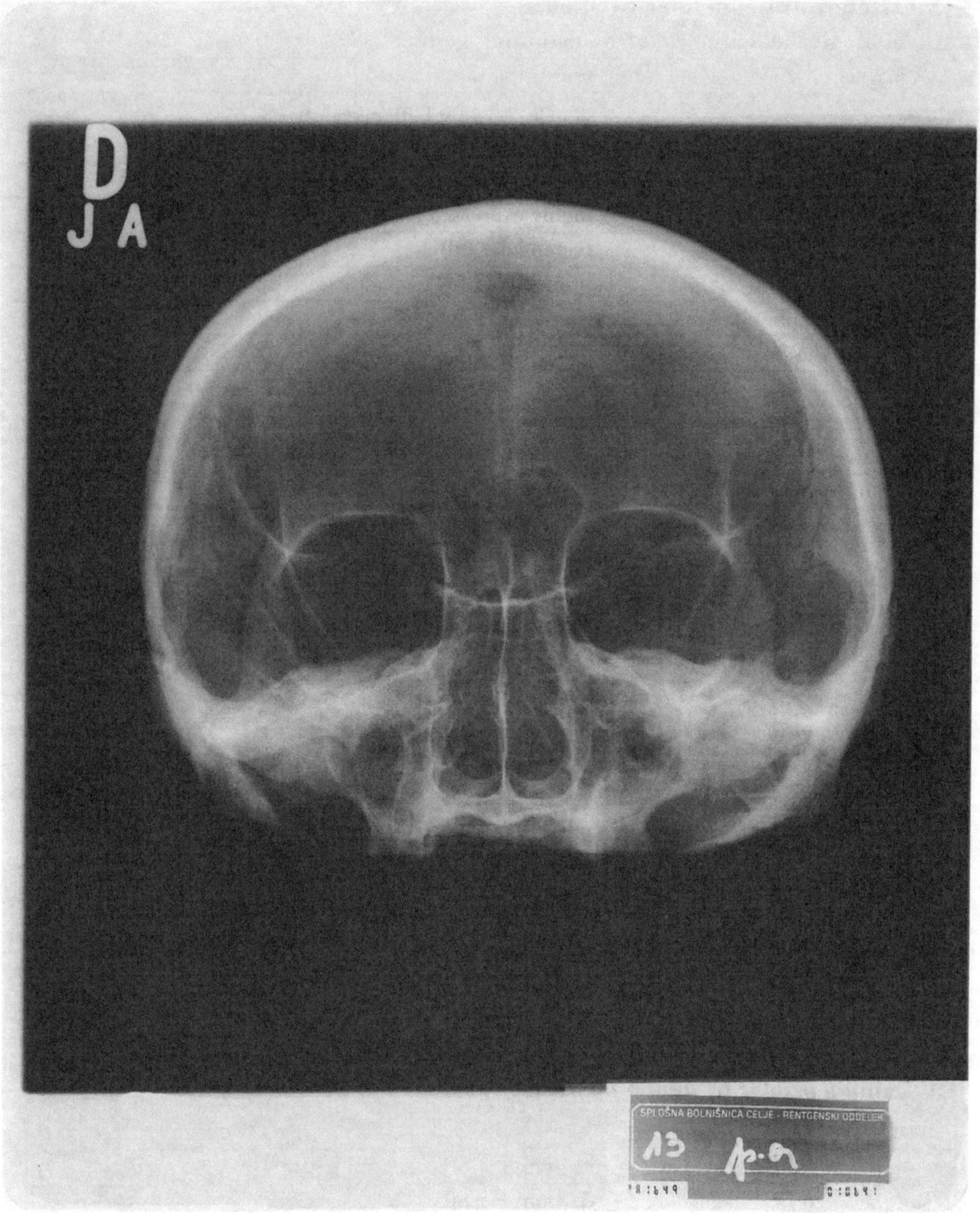

Picture 59: *Occipito-frontal radiograph of skull 13*

4.2.1.13.2. Determining age by skull

Determining age by endo- and ectocranial obliteration of the sutures

The sagittal and coronal sutures are both obliterated, while the lambdoidal is still open in the asterical part. The medial and transversal palatine sutures are partially obliterated. Following the assessment of the state of the skull sutures it may be concluded that the skull belonged to a person aged over 40 years.

Determining age by forensic-dental methods

The alveolar process is protruding by approx. 7 mm, forming a Campers' angle of about 80 degrees. The palate is large, strong and semicircular. The maxilla has a toothless alveolar ridge of semicircular shape. The alveoli of all 16 teeth are preserved. The hard palate is undamaged. On the left side the palatine torus is expressed. The incisive foramen is within normal limits. The medial and transversal palatine sutures are almost closed.

Maxillary alveolar process from right towards left:

	M3	M2	M1	PM2	PM1	C	I2	I1	I1	I2	C	PM1	PM2	M1	M2	M3
Maxilla:	+	+	+	+	+	+	+	+	+	+	+	+	+	+	+	+

Legend: *+ = alveolus present*

Determining age by X-ray methods (pict. 59)

The fronto-occipital radiograph shows a skull with large square orbits, the nasal root is relatively broad, and the nasal septum straight and fully preserved.

On the left and right side views there is a prominent occipital protuberance, as if the cortex had been thickened, but it is only the consequence of the sum of the projections. The whole calvaria is approximately of even thickness, although perhaps slightly thicker occipitally. No other particularities were noted from the radiographs.

no. 13: ♂, around 40 years

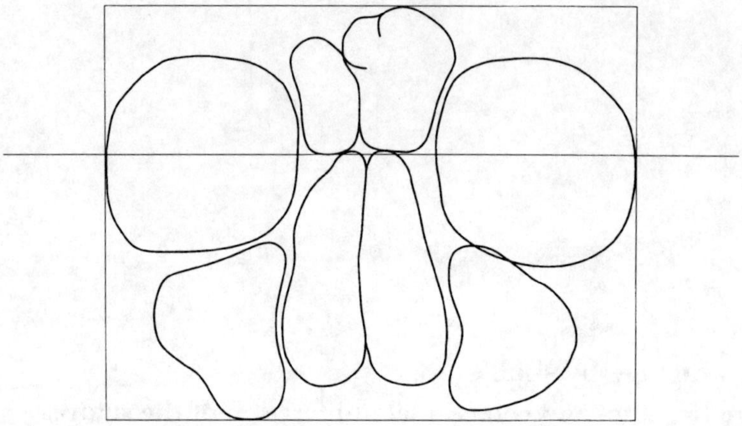

Picture 60: *Outline of the facial cavities of skull 13 from the occipito-frontal radiograph*

4.2.1.13.3 X-ray determination of the facial cavities (pict. 60)

The occipito-frontal radiograph shows that:

- the frontal sinuses are somewhat smaller, asymmetric and leaf shaped;
- the orbits are slightly asymmetric and rectangular. After Nemeskéri they are scored as (+1);
- the maxillary sinuses are relatively symmetrical and triangular or leaf shaped;
- the nasal cavity is symmetrical and similar to the scallop shape with a serrated margin.

Planimetric values of the facial cavities

Skull		Frontal sinuses		Orbits		Maxillary sinuses		Nasal cavity
		Right	Left	Right	Left	Right	Left	
No. 13	S	1.96	6.59	18.61	19.32	7.39	7.48	15.05
	H	1.96	3.30	5.17	4.99	3.74	3.83	5.70
	B	1.16	2.76	4.37	4.45	2.41	2.50	3.30
	T	0.54	0.72	0.81	0.81	0.63	0.54	0.54
	Σ	8.55						
	Sh	0.81	0.63	0.81	0.81	0.72	0.72	0.63

Legend: *S = surface, H = maximal height, B = maximal breadth, T = B/H, Σ = total surface of frontal sinuses, Sh = shape of sinuses*

4.2.1.13.4. Paleopathological methods

The mandibular fossa is smooth on the right, while on the left it is more porous with characters of erosion or degenerative changes. The cranial bones display porosity.

On the left side of the hard palate the palatine torus is expressed. There are no other paleopathological changes on the skull.

4.2.1.13.5. Commentary

The skull is intensively marbled over its entire surface, and according to the cranial index estimate it is short. It is fairly massive and round in shape. On account of posthumous damage to the basilary and mastoidal part of the skull, it is not possible to examine some important features of gender dimorphism, such as the occipital relief (partially preserved) and partial protuberance, and the mastoids. Nevertheless, it can be said that the skull has predominantly masculine traits, from the glabella and supraorbital arches to the inclination of the forehead and the frontal tubers.

The orbits are rectangular. The nasal root is without particular features, the nasal bone partly broken off, and the nasal septum straight and wholly preserved.

The right cheekbone is preserved, the left is missing the cheekbone arch; morphologically, they are without particular features. The palatine ridge of the maxilla is ellipsoid. The dental index after Pont indicates that the alveolar arch is semicircular. It retains all the alveoli preserved; the teeth fell out posthumously. The maxilla is also protruded, the occlusion disgnathic, as is indicated by Campers' angle. The above described features shaped a broad face with healty, but slightly outward protruding teeth.

The left mandibular fossa is porous, the right without particularities. Most of the cranial sutures are obliterated to a considerable extent. The thickness of the bones amounts frontally to 4 mm, occipitally to 5 mm.

All these findings lead to the conclusion that the skull belonged to an adult male. Considering the fairly advanced degree of obliteration of the cranial sutures, it may be conjectured that he was over 40 years of age. The teeth are exceptionally healthy for that age. In comparison with the other skulls in this research, the teeth are outstandingly healthy.

The question of why the skull is marbled remains unexplained.

THE SKULL BELONGED TO AN ADULT MALE, AGED AROUND 40 YEARS

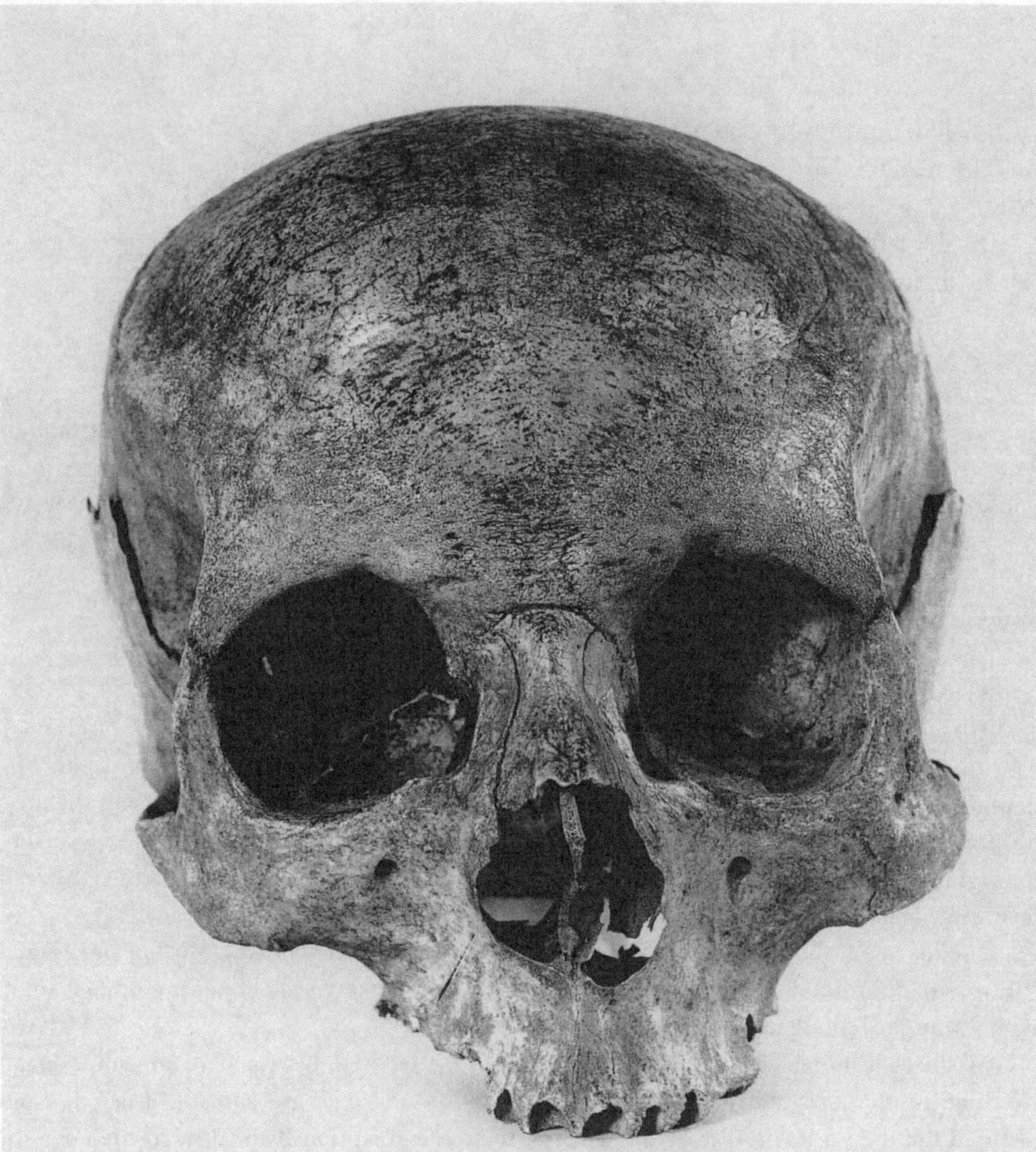

Picture 61: *Skull 14 from the front. Visible is the well expressed asymmetry of the anterior nasal aperture and of the entire maxilla*

4.2.1.14.

Skull 14 (picts. 61,64)

4.2.1.14. Determining gender by skull

Anthropological characters for determining gender

superciliary arches	robust (+2)
glabella	strong (+2)
frontal tubers	robust (+2)
inclination of the forehead	very steep (+2)
supraorbital margin	highly expressed (+2)
orbital shape	rectangular (+2)
occipital relief	strongly pronounced (+2)
ext. occipital protuberance	robust (+2)
mastoid processes	robust (+2)
cheekbones	strong (+2)
maxillary palatine process	asymmetric and deformed
maxillary alveolar process	asymmetric, inclined to right
thickness of calvaria	occipitally greater than frontally
facial part of skull	expressed

Determining gender by anthropometric methods and indices

bizygomatic breadth	after reconstruction 142 mm
great occipital foramen	rounded, length 37 mm, breadth 30.5 mm
mastoids	r. 37 mm, l. partially broken off 28 mm
breadth of anterior nasal aperture	27 mm
cranial length	187 mm
cranial breadth	155 mm
cranial base	110 mm
cranial index	82.89 (short skull)
orbital height	r. 40 mm, l. 35 mm
orbital breadth	r. 35 mm, l. 35 mm
orbital index	left and right – 87.5 (male feature)

Determining gender by dental methods

Dental index after Pont:

Breadth between first premolars	40 mm. P = 50
Breadth between first molars	50 mm. M = 83.33

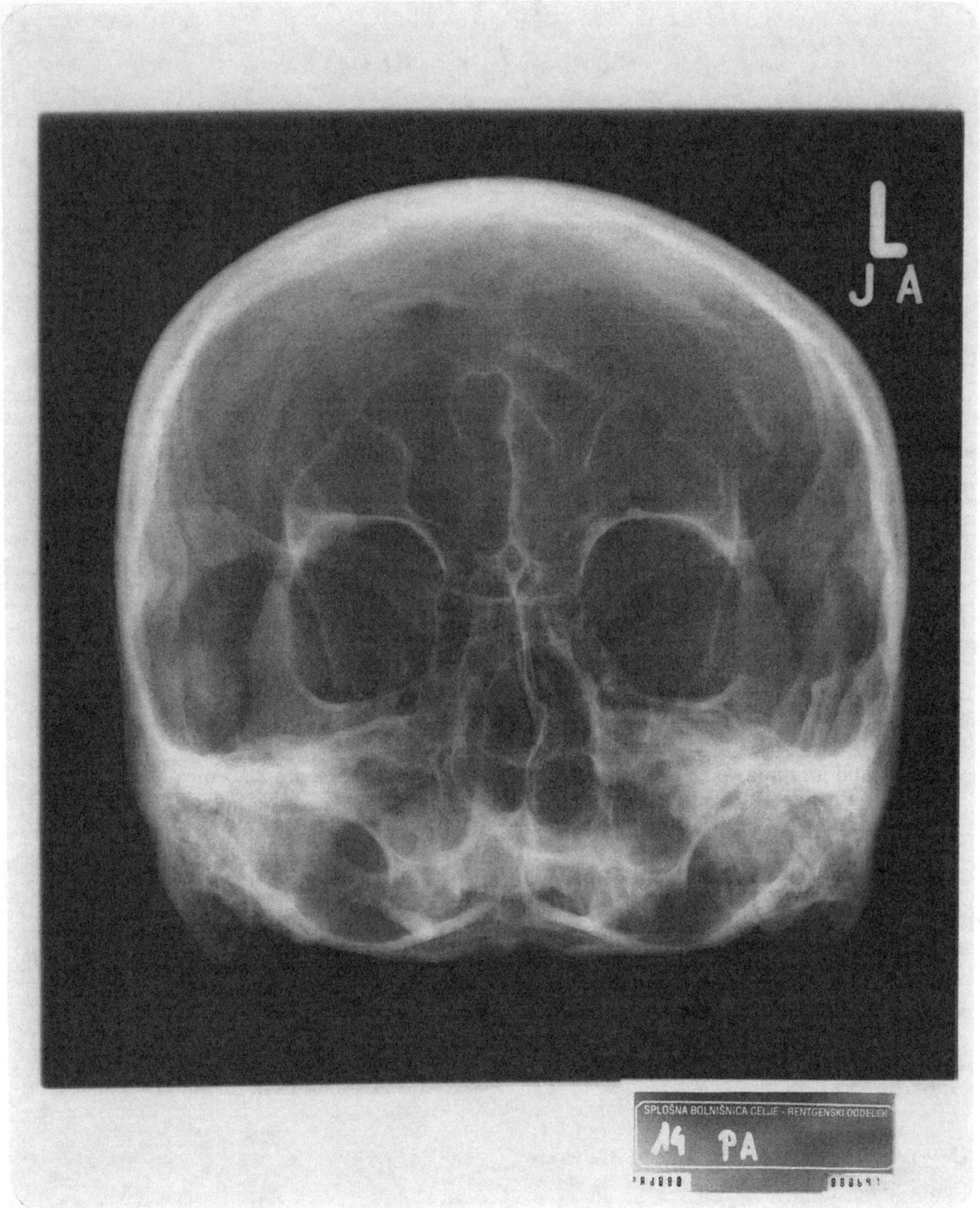

Picture 62: *Occipito-frontal radiograph of skull 14*

4.2.1.14.2. Determining age by skull

Determining age by endo- and ectocranial obliteration of the sutures

The sagittal suture is completely obliterated, the coronal suture has the pars temporalis still open, and the asteric parts of the lambdoidal suture are also not yet obliterated. The medial palatine suture is obliterated. Assessment of the condition of the cranial sutures indicates a person aged around 50 years.

Determining age by forensic-dental methods

The maxilla has a toothless alveolar ridge which, on account of deformation of the palate, is contracted and tilted towards the right. The dental alveoli right and left are preserved. The incisive foramen is preserved and shifted towards the right.

Alveolar maxillary process

Right: a more detailed examination reveals that most of the right dental alveoli are lacking the buccal bone; this is particularly evident on the alveoli of both incisors, the canine and the first premolar. The alveoli of the second premolar and of the first and second molars are shallow, partly atrophied and indicate that these teeth were lost during the lifetime.

The alveoli on the left are better preserved: from 1 to 5 they also have the bone preserved on the buccal side. The alveoli of the molars are missing the buccal bone, and the alveoli are shallower; the person probably lost these teeth during lifetime.

Maxillary alveolar process from right towards left:

	M3	M2	M1	PM2	PM1	C	I2	I1	I1	I2	C	PM1	PM2	M1	M2	M3
Maxilla:	–	+*	+*	+*	+	+	+	+	+	+	+	+	+	+**	+*	–

Legend: *+* = shallow alveolus, +** = distal root preserved*

Determining age by X-ray methods (pict. 62)

The radiographs show a robust skull. On the fronto-occipital view one may observe rather rectangular orbits, the nasal bone is relatively broad, with the septum deviating to the left. Parietally there are visible stains, which belong to Pacchionian bodies.

On the side views the frontal bone at the height of the sinuses is very thick (hyperostosis frontalis interna). The maxilla is retracted. Occipitally the bone is thicker than frontally. The grooves of the meningeal artery are within normal limits.

Dorso-parietally one may see a solitary formation, the size of an almond, with relatively sharp margins. By using differential diagnostics an attempt to distinguish between a physiological and a pathological formation was made. Physiologically, it could be a Pacchionian body, situated more dorso-parietally (Grashey, Birkner 1964). Among the pathological variants a cystic or chronic inflammatory process, or less probably an infiltrating process was considered. In the case of osteolysis the margins are usually non-sharp and non-sclerotized, while in metastases they are sharper. Detailed analysis indicated a dorso-parietal variant of a Pacchionian body.

The remaining radiographs were within normal limits.

no. 14: ♂, around 40 years

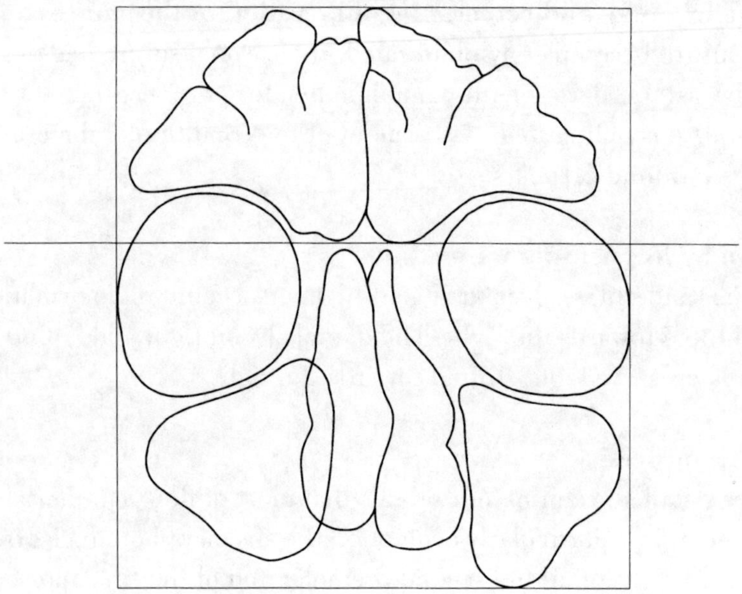

Picture 63: *Outline of the facial cavities of skull 14 from the occipito-frontal radiograph*

4.2.1.14.3. X-ray determination of the facial cavities (pict. 63)

The occipito-frontal radiograph shows that:

- the frontal sinuses are extremely large, somewhat asymmetric and deeply septicized. In shape they are large fanlike-pyramidal (see theoretical proposal). Comparable sinuses are presented by the X-ray atlases as so-called pneumosinus dilatans (Grashey, Birkner 1964);
- the orbits are symmetrical and rectangular. After Nemeskéri they are scored as (+2);
- the maxillary sinuses are relatively symmetrical. They resemble a triangular shape;
- the nasal cavity is non-symmetrical. It is scallop shaped with serrated margins.

Planimetric values of the facial cavities

Skull		Frontal sinuses		Orbits		Maxillary sinuses		Nasal cavity
		Right	Left	Right	Left	Right	Left	
No. 14	S	20.66	20.66	17.26	16.66	7.48	7.57	16.66
	H	6.12	6.46	5.26	4.68	3.49	4.25	6.97
	B	5.02	5.02	4.17	4.00	2.64	2.21	3.32
	T	0.68	0.68	0.77	0.77	0.68	0.43	0.43
	Σ	40.72						
	Sh	0.60	0.43	0.77	0.85	0.68	0.60	0.51

Legend: *S = surface, H = maximal height, B = maximal breadth, T = B/H, Σ = total surface of frontal sinuses, Sh = shape of sinuses*

4.2.1.14.4. Paleopathological methods

In the case of skull 14, the maxillary palatine process is the most dramatically changed. All the anatomically present structures are obliterated. One cannot see either the medial palatine suture nor the palatine suture transverse, and the palatine foramens are not visible as well. The maxillary palatine process is completely asymmetric, the right being far more deformed than the left and its surface is uneven. It is covered all over with minute perforations which are indicative of resorptive changes, with minor convergences of these occurring at the centre.

The maxillary alveolar process is also strongly changed. The left dental alveoli are relatively well preserved at the incisors, the canine and first molars, and indicate posthumous loss of teeth, while the alveoli of the left back molars are damaged. The right alveolar arch is strongly changed: the dental alveoli are preserved only for the incisors and the canine; the remaining interdental septa are missing, and the partially preserved atrophic, but badly damaged alveoli can be seen.

The pathological changes are proliferative and scleroticizing, and have spread diffusely into the hard palate, the alveolar excrescence of the maxilla, and into the nasal bone, the septum and surroundings. The structures mentioned are altered in shape and disfigured. The anterior nasal aperture is completely asymmetric, the entire central part of the face is deformed and shifted to the right. The finely granulated osseous growth is most widely spread in the hard palate. On account of the growing process, the person not only lost the upper right tooth, but must also have had a face of monstrous appearance. In addition, this skull is enormously large, and therefore our conjectures were also directed towards acromegaly. Following thorough studies and numerous consultations with subspecialists, the diagnosis of Albright's syndrome was reached, the rarest form of fibrous dysplasia, which is linked to acromegaly. This is described in 3.2.3.1. under paleopathological methods.

The following were also considered in the differential diagnostics:
- osteosarcomas
- osteomyelitis
- fibrous dysplasia
- gigantocellular granuloma,
- eosinophilic granuloma.

The mandibular fossae on both sides are without special features. No other paleopathological characters were observed on skull 14.

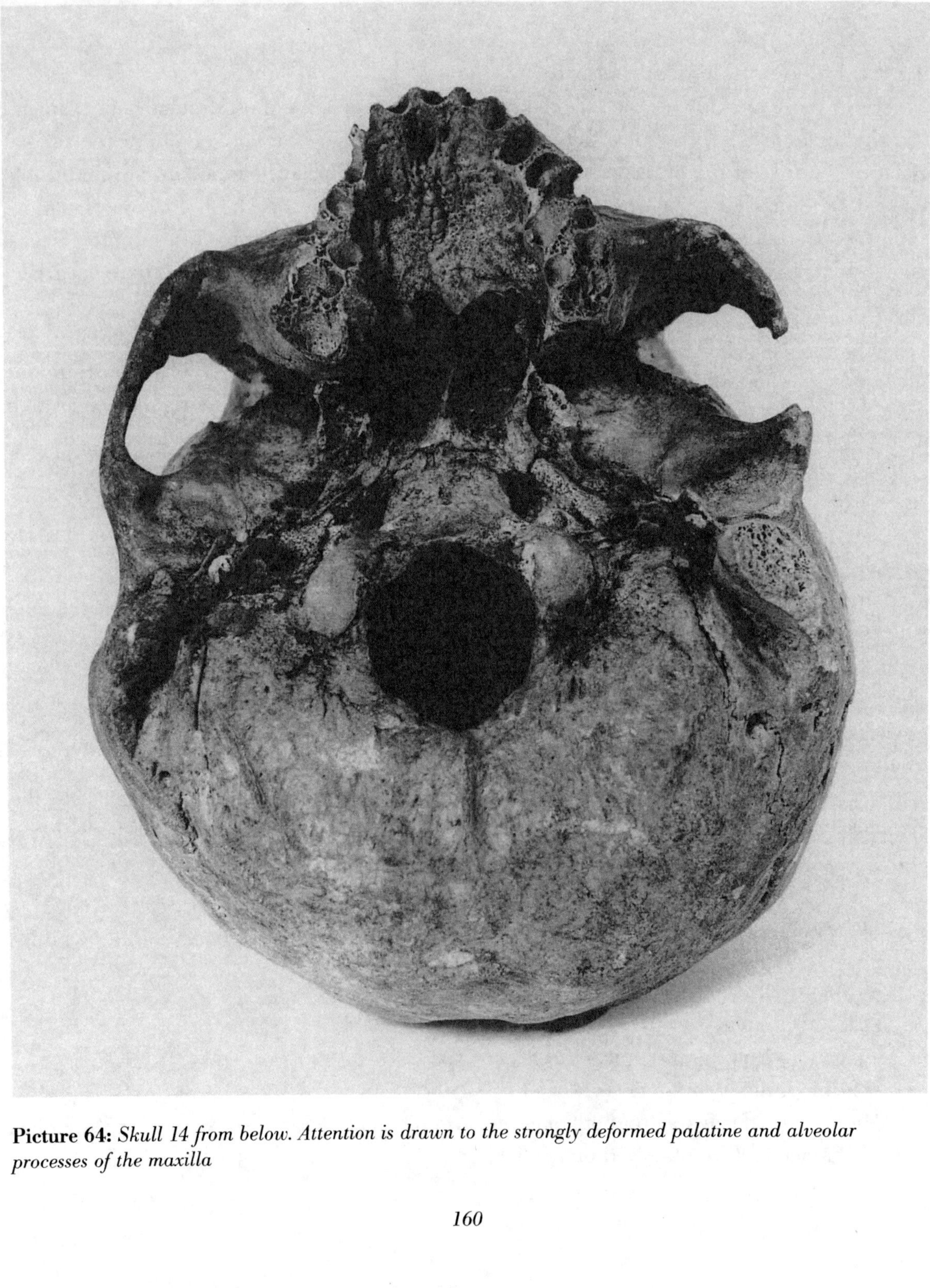

Picture 64: *Skull 14 from below. Attention is drawn to the strongly deformed palatine and alveolar processes of the maxilla*

4.2.1.14.5. Commentary

Skull 14 is the most robust among all the skulls studied. It displays well expressed masculine characters and has a highly deformed central part of the face. According to the cranial index it is short, generally round in shape and slightly damaged.

Among the typical masculine characters, one notes on this skull: the expressed glabella and supraorbital arches, the strongly pronounced frontal and parietal tubers, and the steeply inclined forehead. Likewise on the occiput one notes the strong occipital protuberance and the muscle attachments on it, the large mastoids, and the rounded occipital opening. The strong nasal bone with the septum curving to the left, the expressed cheekbones and alveolar ridge of the maxilla give evidence that this skull had a robust face. The orbits are more rectangular in shape. Campers' angle indicates that the maxilla is protuberant, and hence the facial appearance even more robust.

The rare pathology discovered on this skull is described in the section above (paleopathology).

This person, with such strong facial asymmetry, would have stood out on account of his appearance. No mention of this could be traced in the historical records.

THE SKULL BELONGED TO AN ADULT MALE, AGED AROUND 40 YEARS

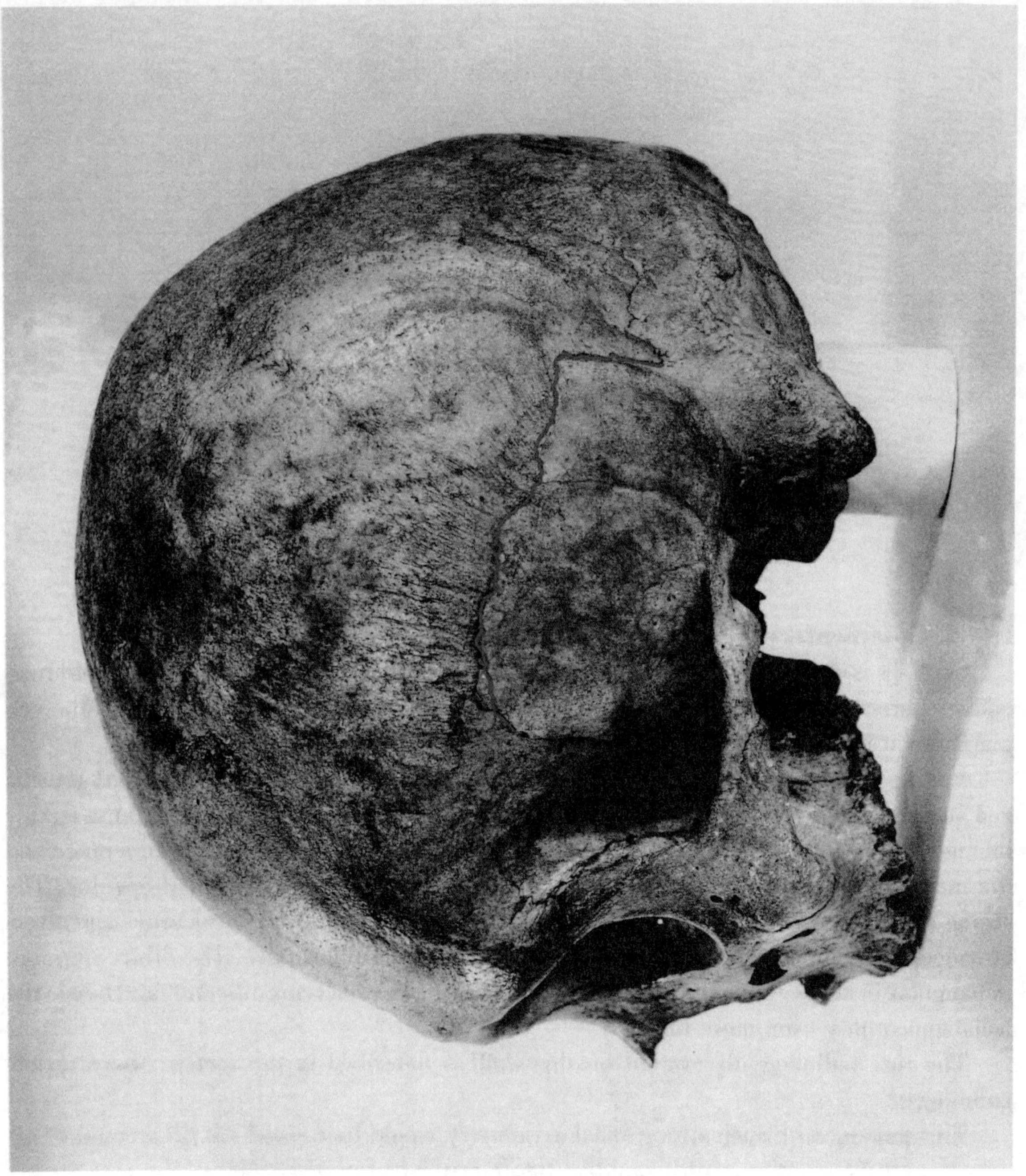

Picture 65: *Skull 15 from the side: the temporal bones left and right have a relatively even upper rim, comparable to the temporal bones of skull 17*

4.2.1.15.

Skull 15 (pict. 65)

4.2.1.15.1. Determining gender by skull

Anthropological characters for determining gender

superciliary arches	strongly developed (+2)
glabella	strong (+2)
frontal tubers	strongly expressed (+2)
inclination of the forehead	steep (+2)
supraorbital margin	highly expressed (+2)
orbital shape	trapezoid (+2)
occipital relief	pronounced (+2)
ext. occipital protuberance	moderately developed (+1)
mastoid processes	very strong (+2)
cheekbones	moderately developed (+1)
maxillary palatine process	semicircular
maxillary alveolar process	semicircular
thickness of calvaria	occipitally greater than frontally
facial part of skull	expressed

Determining gender by anthropometric methods and indices

bizygomatic breadth	135 mm
orbital height	r. 35 mm, l. 35 mm
orbital breadth	r. 42 mm, l. 42 mm
great occipital foramen	rounded, length 34 mm, breadth 30 mm
mastoids	both preserved, l. 40 mm, r. 28 mm
breadth of anterior nasal aperture	30 mm
cranial length	179 mm
cranial breadth	143 mm
cranial base	107 mm
cranial index	79.89 (medium length skull)
orbital index	r. and l. 83.33. Value suitable for a male skull.

Determining gender by dental methods

Dental index after Pont:

Breadth between first premolars	43 mm. P = 53.75
Breadth between first molars	52 mm. M = 86.66

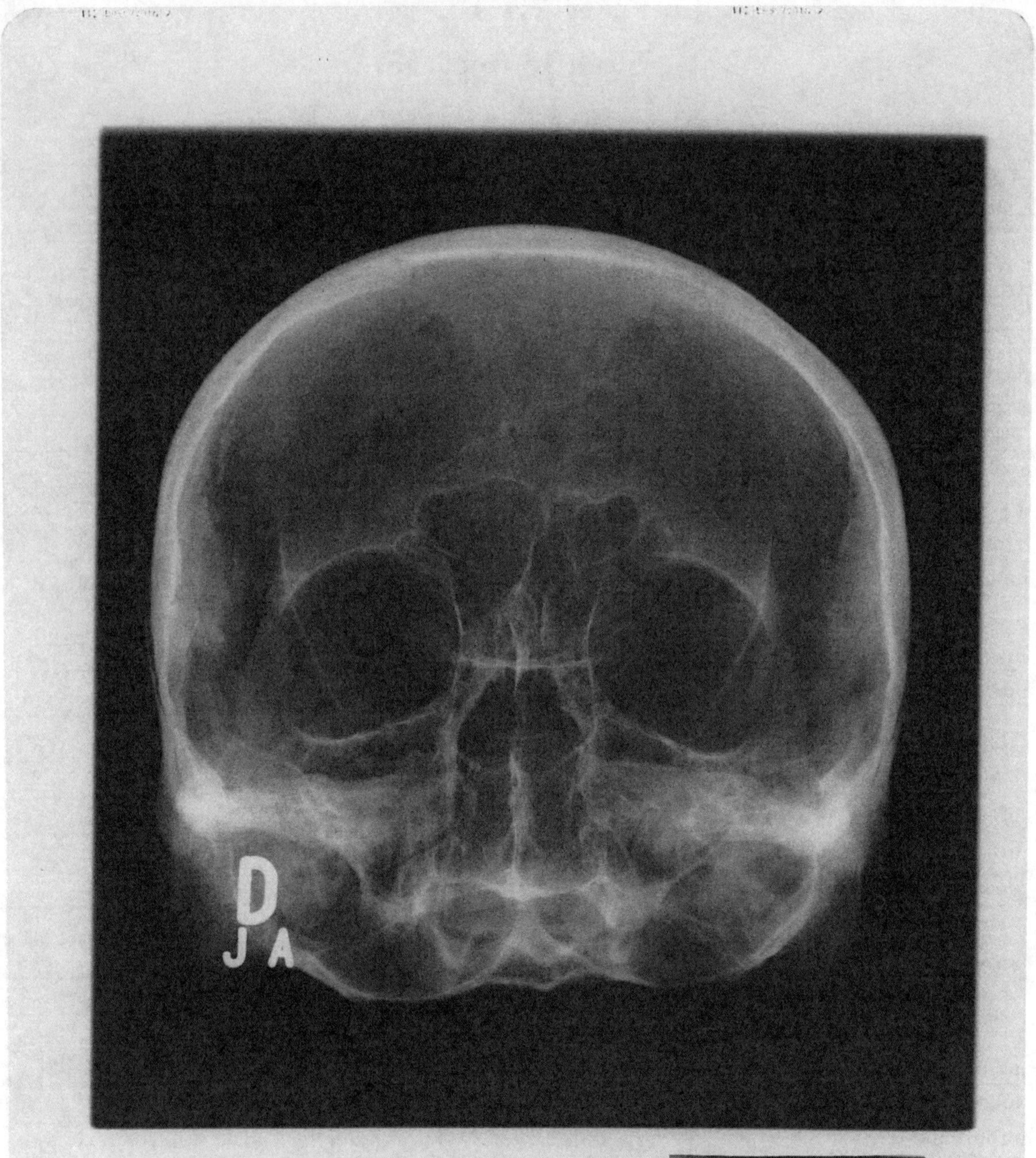

Picture 66: *Occipito-frontal radiograph of skull 15*

4.2.1.15.2. Determining age by skull

Determining age by endo- and ectocranial obliteration of the sutures

The sutures are well seen; the sagittal and coronal sutures are mainly obliterated, while the lambdoidal still has the asterical part open. The sutures on the hard palate, the medial and transversal, are already to a considerable extent obliterated.

Determining age by forensic-dental methods

The maxilla is toothless. The alveolar arch is semicircular and, together with the hard palate, completely preserved. The elliptically shaped palatine torus runs along the medial palatine suture on both sides; it ends at the front in the incisive foramen. The small person must have had good teeth and retained the majority of them until the end of his lifetime. Right: the alveolus of the first incisor is atrophic, the alveoli of the remaining teeth right (2–7) are deep and indicate that they fell out after death. There is no third molar; this may be hypodontia.

Left: the alveoli of eight teeth. The front alveoli are deep, indicating posthumous loss of teeth; the alveoli of the molars are somewhat shallower and also partially damaged.

Maxillary alveolar process from right towards left:

	M3	M2	M1	PM2	PM1	C	I2	I1	I1	I2	C	PM1	PM2	M1	M2	M3
Maxilla:	–	+	+	+	+	+	+	a	+	+	+	+	+	+	+*	+*

Legend: *a = atrophic, +* = partially preserved alveolus*

Determining age by X-ray methods (pict. 66)

The fronto-occipital projection shows a massive skull with trapezoidal orbits; the nasal bone is strong and long, with the septum deviating slightly to the right. The Pacchionian bodies are within the physiological boundaries. The right mastoid is larger than the left. The skull (calvaria) is thicker occipitally than frontally. The sagittal suture is obliterated, but the lambdoidal is still partially open.

On the side views left and right the well expressed grooves of the meningeal artery and the diploic vein are visible. The septa between the dental alveoli of the maxilla are well preserved on both sides. Likewise, on both sides the supraciliary arches are both pronounced.

The pictures of the base are without particularities. There are also no other special features to be noted on the radiographs.

no. 15: ♂, 30–40 years

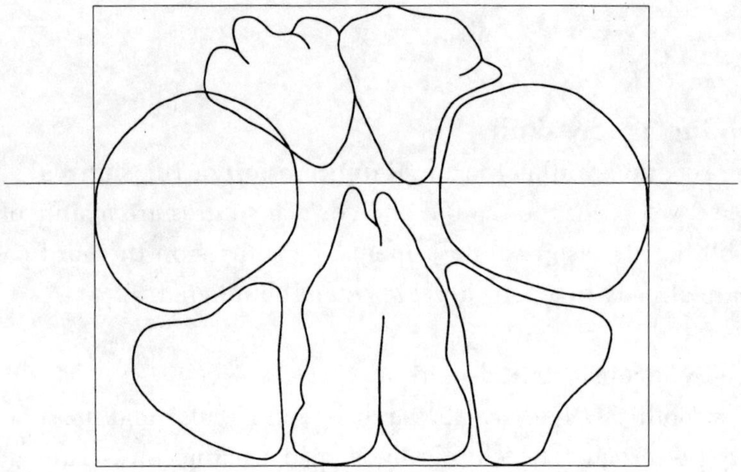

Picture 67: *Outline of the facial cavities of skull 15 from the occipito-frontal radiograph*

4.2.1.15.3. X-ray determination of the facial cavities (pict. 67)

The occipito-frontal radiograph shows that:

- the frontal sinuses are of medium size, non-symmetrical and fan shaped;
- the orbits are between rectangular and trapezoid in shape. After Nemeskéri they are scored as (+2);
- the maxillary sinuses are fairly symmetrical and triangular in shape;
- the nasal cavity is also symmetrical and scallop shaped with a serrated margin.

Planimetric values of the facial cavities

Skull		Frontal sinuses		Orbits		Maxillary sinuses		Nasal cavity
		Right	Left	Right	Left	Right	Left	
No. 15	S	8.64	7.30	18.34	17.18	6.15	6.23	17.72
	H	3.83	3.56	4.99	4.72	3.12	3.30	6.68
	B	3.21	2.76	4.45	4.19	2.32	2.23	3.83
	T	0.72	0.72	0.81	0.81	0.63	0.63	0.54
	Σ	15.94						
	Sh	0.72	0.63	0.89	0.81	0.81	0.81	0.45

Legend: *S = surface, H = maximal height, B = maximal breadth, T = B/H, Σ = total surface of frontal sinuses, Sh = shape of sinuses*

4.2.1.15.4. Paleopathological methods

Articulating surfaces of both temporomandibular joints are finely porous.

The person had very healthy teeth; to the outside observer it must have been evident that he had lost the first right incisor at an early age.

On skull 15 no other pathological characters are preserved.

4.2.1.15.5. Commentary

According to the cranial index, this skull is scored as medium long. It is rounded in shape, robust and almost undamaged. It is a most typical masculine skull with exceptionally pronounced traits on the front part of the skull: the glabella, supraorbital arches, frontal and parietal tubers, steep forehead. The occipital features also substantiate the conclusion that this is a male skull: the occipital protuberance is strong in relief, so are the large mastoids and the more rounded occipital openings. The face is shaped by expressed orbits, which according to the orbital index are grouped among typically masculine traits; the cheekbones are also robust, and Campers' angle indicates a balanced occlusion. According to the dental index after Pont, the palatine maxillary process is semicircular, so too is the alveolar process. In the dentition almost all the dental alveoli are preserved, with the exception of the atrophic alveoli of the right first incisor, which was lost during the person's lifetime. Otherwise, the teeth were healthy and were retained until the end of his life. They fell out posthumously. The above described features formed a rather broad, harmoniously structured face, and the nasal bone also contributed to its appearance.

The skull reveals no pathological changes, apart from the wearing down detected on the temporomandibular joints. The cranial sutures are obliterated, thus indicating an adult person aged around 40 years.

It may, therefore, be concluded that the skull belonged to a healthy adult male. The cause(s) of death cannot be determined from the skull.

THE SKULL BELONGED TO AN ADULT MALE, AGED 30–40 YEARS

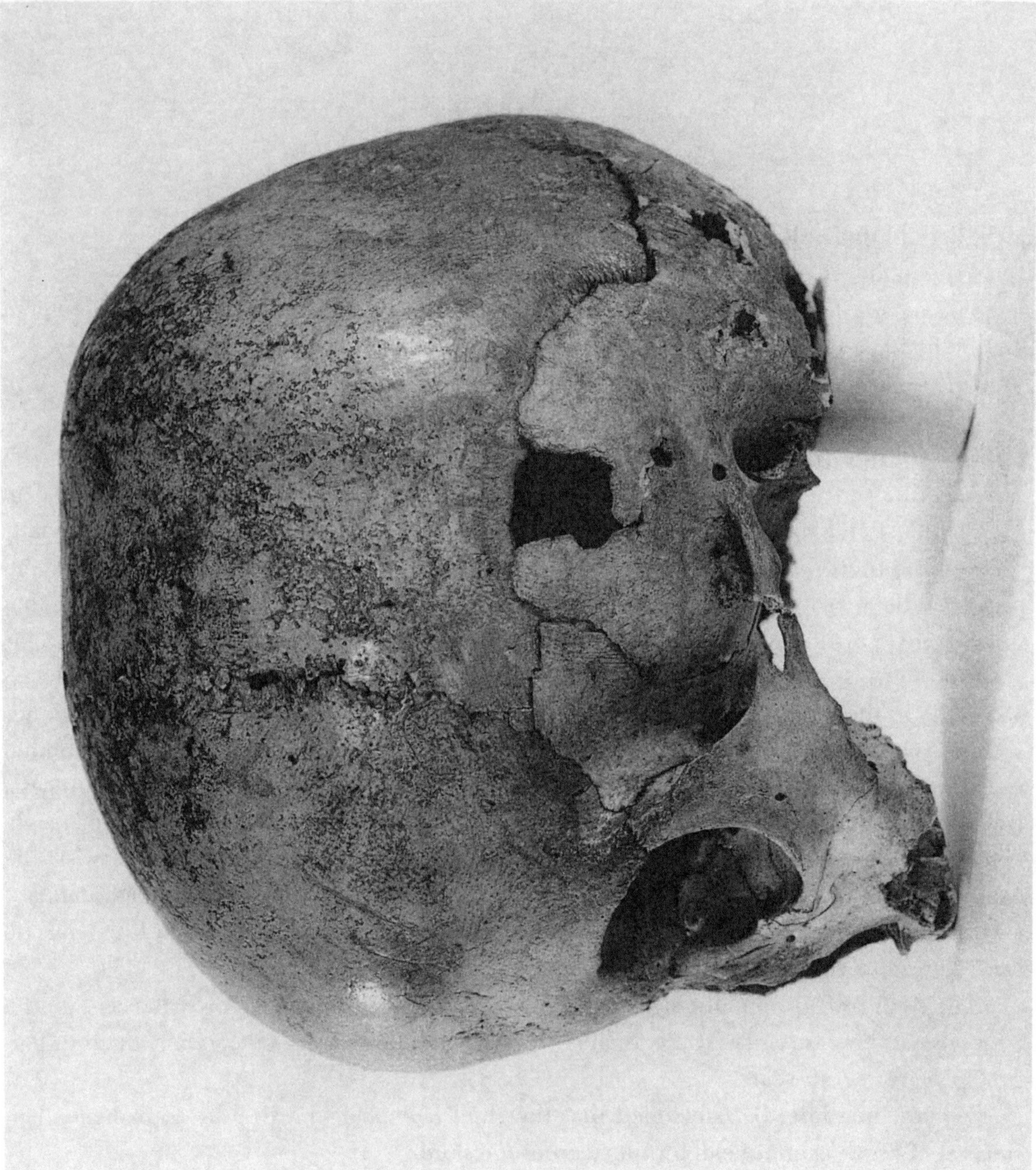

Picture 68: *Side view of a tiny and considerably damaged child's skull, marked as no. 16*

4.2.1.16
Skull 16 (pict. 68)

4.2.1.16.1 Determining gender by skull

Anthropological characters on the skull for determining gender

superciliary arches	undeveloped (–2)
glabella	nonexistant (–2)
frontal tubers	slightly developed on both sides (–2)
inclination of the forehead	slightly inclined forwards (–2)
orbital shape	rounded (–2)
supraorbital margin	thin
cheekbones	r. partly broken off, left gracile (–2)
occipital relief	poorly developed (–2)
ext. occipital protuberances	undeveloped (–2)
mastoid processes	slight (–2)
maxillary palatine process	ellipsoid
maxillary alveolar process	semicircular
thickness of calvaria	very thin throughout
facial part of skull	small and narrow

Determining gender by anthroscopic and anthropometric methods and indices

bizygomatic breadth	after reconstruction 111 mm
great occipital foramen	broken off
orbital height	r. 32 mm, l. 32 mm
orbital breadth	r, 34 mm, l. 34 mm
mastoids	r. missing entirely, l. missing tip,
	measuring in breadth 20 mm, in length 35 mm
breadth of anterior nasal aperture	24 mm
cranial length	168 mm
cranial breadth	146 mm
cranial base	undeterminable due to damage
Campers' angle	almost 90 degrees
cranial index	86.90 (very short skull)
orbital index	r. and l. 94.12 (female feature)

Determining gender by dental methods

Dental index after Pont:

Breadth between first premolars	32 mm. P = 40
Breadth between first molars	41 mm. M = 68.33

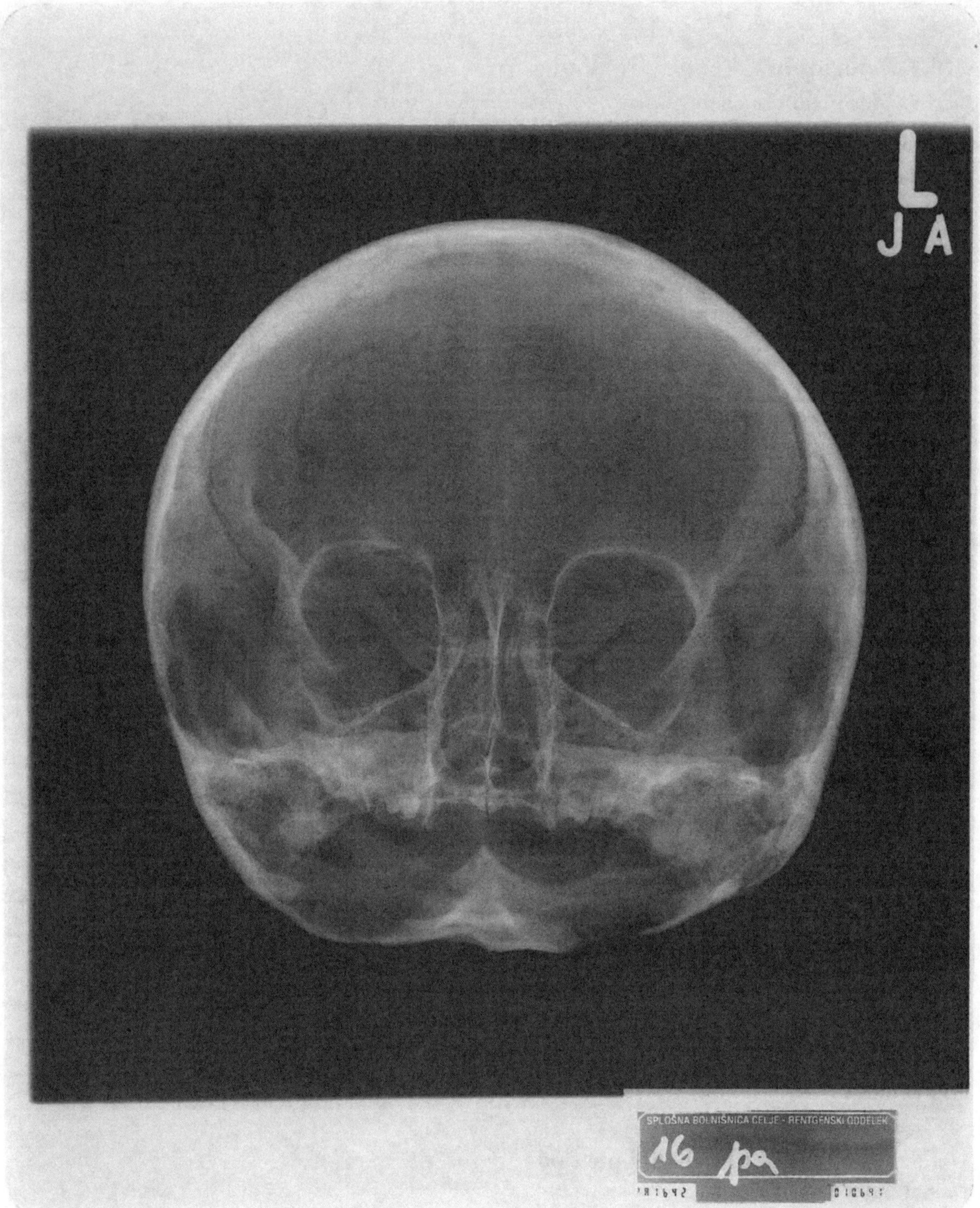

Picture 69: *Occipito-frontal radiograph of skull 16*

4.2.1.16.2. Determining age by skull

Determining age by endo- and ectocranial obliteration of the sutures

The spheno-occipital synchondrosis, which indicates a child's skull, is open. The sagittal, coronal and lambdoidal sutures are all wide open. The same is true for the medial palatine suture and the transversal, and also the temporozygomatic left and right. The temporoparietal suture is also wide open.

The state of the cranial sutures is indicative of a child's skull.

Determining age by forensic-dental methods

The maxillary alveolar arch is small and semicircular in shape; the alveolar process is symmetrical and broken off on both sides at the height of the first molar. Also missing is the pharingeal part of the palatine process of the maxilla. The left alveolar arch has a considerable part of the labial side broken off, while the buccal side is preserved. The exchange of milk teeth for permanent ones is seen.

The medial and transversal palatine sutures are both open. The incisive foramen is preserved. Only one tooth remains preserved.

Maxillary alveolar process from right towards left:

	M3	M2	M1	PM2	PM1	C	I2	I1	I1	I2	C	PM1	PM2	M1	M2	M3
Maxilla:	–	r	+	mr	mr	mr	+	+	+	+	mr*	t	mr	+	r*	–

Legend: *t = tooth, r = retained tooth, r* = retained tooth fell out, mr = alveolus of milk tooth and retained tooth*

(Left: 1 2 III (3) tooth (4) V (retained 5) 6 (7, retained, fell out?) (8, no base yet)
(Right: 1 2 III (retained 3) IV (retained 4) V (retained 5) 6 (retained 7, although tooth fell out) (8, no base yet)

Right, permanent teeth have grown out: both incisors, first molar. Still retained are: canine, first premolar, second premolar. On the radiograph and on the orthopan the still unerupted permanent teeth are visible in the bone: canine, both premolars and second molar. The third molar had not developed yet.

Left, permanent teeth have grown out: both incisors, canine (not retained, probably fell out), first molar.

Still retained: first premolar (probably fell out), second premolar, second molar (probably fell out).

The state of the teeth indicates a child, aged approx. 10–12 years.

no. 16: ♂, 10 years

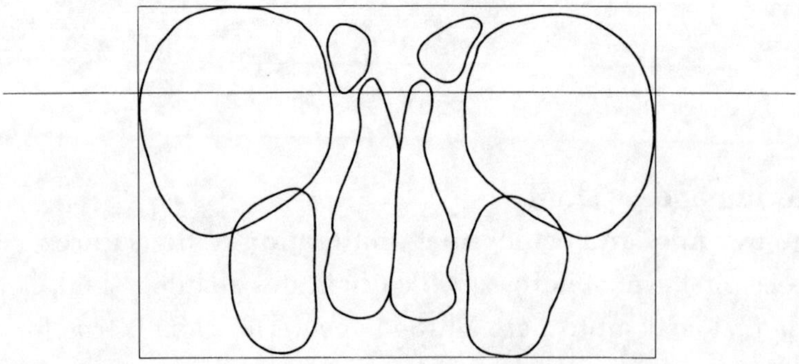

Picture 70: *Outline of facial cavities of skull 16 from occipito-frontal radiograph*

Determining age by X-ray methods (pict. 69)

The radiographs reveal a gracile bone structure with large osseous defects in the occiput (artefact). The calvaria is frontally and occipitally of approximately uniform thickness. Left and right are the pronounced grooves of both meningeal arteries and the diploic veins. There are pronounced digital impressions occipitally.

On the fronto-occipital views the orbits are small and rather rounded, the nasal bone is narrow with an intact nasal septum. The frontal sinuses have not yet developed. All the sutures are open; the sagittal, coronal and the lambdoidal.

Basal view: the expressed openness of the suture between the sphenoid and occipital bones – spheno-occipital synchondrosis is apparent. The lambdoidal suture is also prominently open. In the maxillary alveolar process the still unerupted teeth are visible: right canine, first right premolar, second right premolar, second left premolar.

The preserved alveoli of the first molars left and right convincingly indicate that these two teeth had already grown out and that the child must have been at least 7–9 years old, most probably around 10.

The other radiographs are without particularities.

An orthopan was also made, although it is primarily intended for intravital scanning. Even the shortest possible exposures were still too long for technically suitable pictures. Classical radiographs of skull 16 provided good quality and appropriate answers also in the case of the teeth.

4.2.1.16.3 X-ray determination of the facial cavities (pict. 70)

The occipito-frontal radiograph shows that:

- the frontal sinuses are very small, still undeveloped. With respect to the early developmental phase, they display a bean-like shape, and would most probably have developed into a different shape (considering the state attained, they would be expected to be fan shaped).

- the orbits are more rounded in shape. After Nemeskéri they are scored as (–2);
- the maxillary sinuses are relatively symmetrical. In this developmental phase they are kidney shape, however it is questionable whether they would have remained like that.
- the nasal cavity is symmetrical. It is scallop shaped with a serrated margin.

Planimetric values of the facial cavities

Skull		Frontal sinuses		Orbits		Maxillary sinuses		Nasal cavity
		Right	Left	Right	Left	Right	Left	
No. 16	S	0.89	0.89	15.58	14.69	4.90	4.90	10.77
	H	1.25	1.43	4.54	4.45	2.67	2.67	4.90
	B	0.89	0.81	3.74	3.56	2.32	2.23	3.03
	T	0.63	0.54	0.72	0.72	0.81	0.72	0.54
	Σ	1.78						
	Sh	0.72	0.63	0.85	0.81	0.81	0.72	0.54

Legend: *S = surface, H = maximal height, B = maximal breadth, T = B/H, Σ = total surface of frontal sinuses, Sh = shape of sinuses*

4.2.1.16.4. Paleopathological methods

The mandibular fossa has completely smooth surfaces on both sides.

There are no paleopathological features on skull 16. From the skull alone it is not possible to conclude about the cause of death of this child.

4.2.1.16.5. Commentary

According to the cranial index, this skull is scored as very short. It is gracile and light, ovaloid, and was damaged posthumously. All the characters are indicative of a child's skull in which the gender dimorphism is not yet sufficiently developed for the sex to be determined with certainty. It is possible that the features are more indicative of the female sex. The orbits are round shaped, there is no glabella, the superciliary arches are undeveloped, and the frontal tuber present. The right mastoid is missing in its entirety, the left one is lacking only its peak. The great occipital foramen and the occipital protuberance are both broken off. The entire occipital bone is thin and rough, since the lamina externa of the calvaria has peeled off. The spheno-occipital synchondrosis is mainly open. All the cranial sutures are still open, also at the bottom of the skull. The dental arch is semicircular; within it the milk teeth are being replaced by the permanent teeth. On the right, both incisors and the first molar have grown out; similarly on the left, with the exception of the canine for which, on account of the damage, it cannot be said if it is a permanent tooth already. Only one tooth is preserved: the first left premolar.

There are no disease caused changes.

THE SKULL BELONGED TO A CHILD, AGED APPROXIMATELY 10 YEARS

Gender dimorphism is poorly expressed.

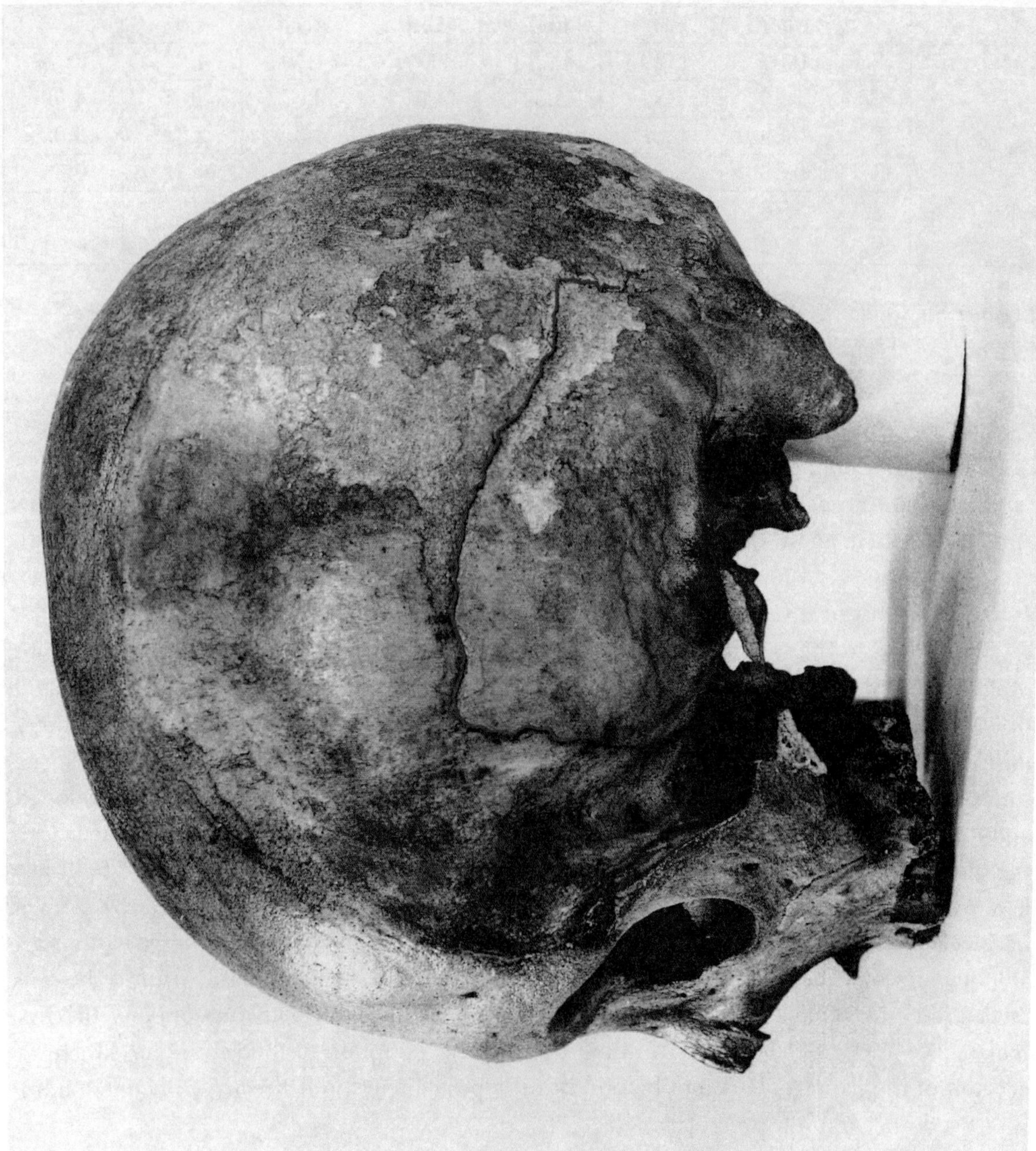

Picture 71: *Skull 17, side view: the temporal bones left and right have a relatively straight upper margin; they are comparable to the temporal bones of skull 15*

4.2.1.17.

Skull 17 (pict. 72)

4.2.1.17.1 Determining gender by skull

Anthropological characters for determining gender

superciliary arches	moderately developed (+1)
glabella	moderate (+l)
frontal tubers	mediumly expressed (+l)
inclination of the forehead	moderately steep (+l)
supraorbital margin	expressed (+1)
orbital shape	r. rectangular, l. trapezoid (+l)
occipital relief	moderately pronounced (+1)
ext. occipital protuberances	moderate (+1)
mastoid processes	expressed (+2)
cheekbones	moderately developed (+1)
maxillary palatine process	relatively narrow, ellipsoid
maxillary alveolar process	semicircular
thickness of calvaria	occipitally greater than frontally
facial part of skull	pronounced

Determining gender by anthropometric methods and indices

bizygomatic breadth	
after reconstruction	broken off
orbital height	r. 34 mm, l. 34 mm
orbital breadth	r. 40 mm, l. 40 mm
great occipital foramen	rounded, length 33 mm, breadth 29 mm
mastoids	both preserved, r. 30 mm, l. 22 mm
breadth of anterior nasal aperture	28 mm
cranial length	176 mm
cranial breadth	141 mm
cranial base	102 mm
cranial index	80.11 (medium length skull)
orbital index	right and left – 85 (conforms to male skull values)

Determining gender by dental methods

Dental index after Pont:

Breadth between first premolars	30 mm. P = 47.5
Breadth between first molars	50 mm. M = 83.33

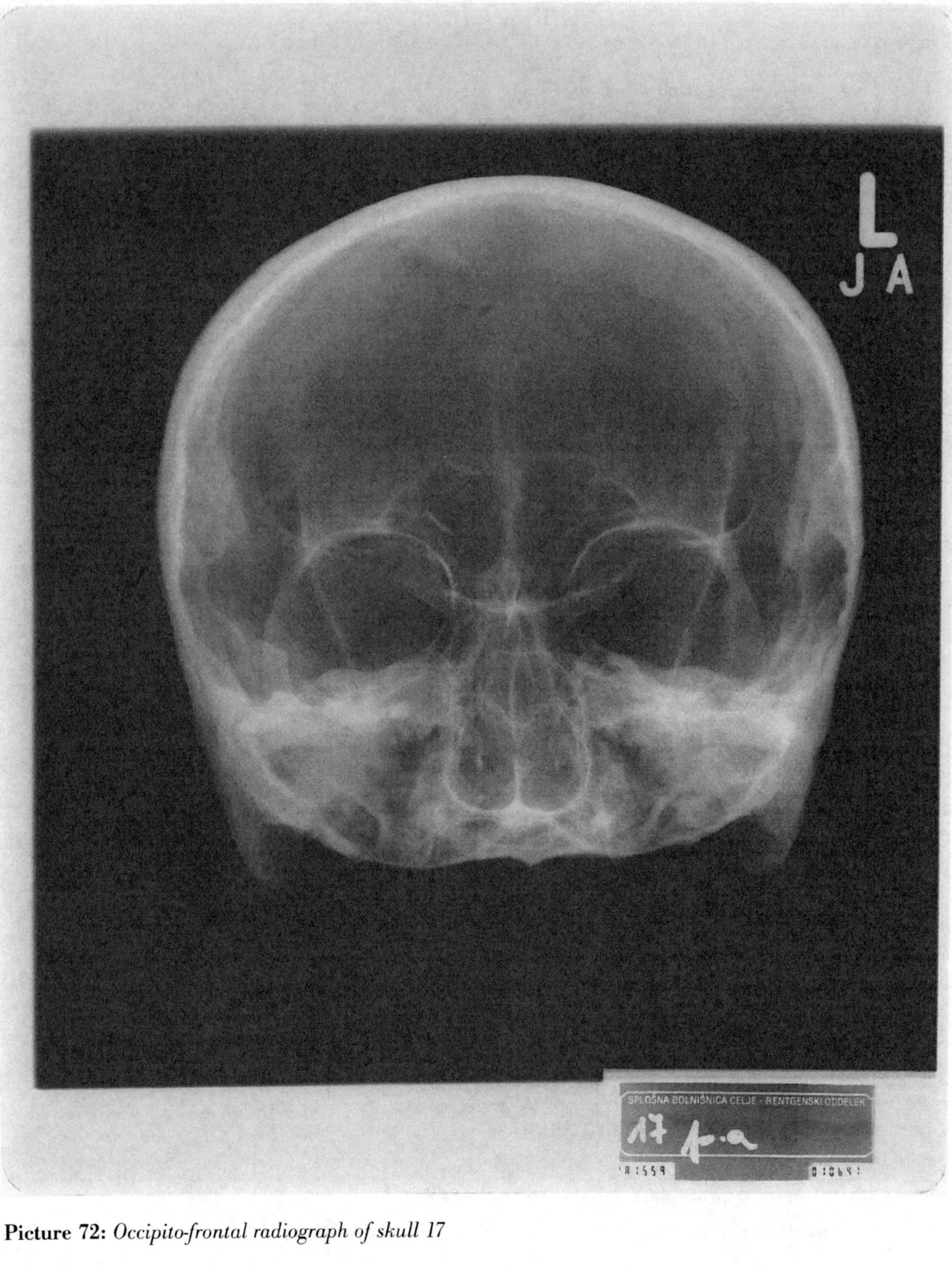

Picture 72: *Occipito-frontal radiograph of skull 17*

4.2.1.17.2. Determining age by skull

Determining age by endo- and ectocranial obliteration of the sutures

All sutures are completely obliterated and are indicative of an old man.

Determining age by forensic-dental methods

The maxillary alveolar process is semicircular, symmetrical, almost undamaged and without teeth. The hard palate has anatomically running sutures. The palatine torus, running on both sides parallel to the medial palatine suture, is expressed. The incisive foramen is large.

Maxillary alveolar process right: the dental alveoli are mainly deep, and indicate that the person lost the teeth after death. Only the alveolus of the first premolar is more shallow. The labial side of the alveolar septa is broken off at the second incisor, the canine and the first premolar.

The maxillary alveolar process on the left is poorly preserved. Some of the alveoli are rather shallow, and most of them are missing the alveolar septa. The shallow atrophic alveoli of both premolars left led to the conclusion that the person's teeth were lost during lifetime.

Maxillary alveolar process from right towards left:

	M3	M2	M1	PM2	PM1	C	I2	I1	I1	I2	C	PM1	PM2	M1	M2	M3
Maxilla:	+	+	+	+	+*	+	+	+	+	+	+	+*	+*	+	+	b

Legend: *+* = shallow alveolus, b = broken off*

Determining age by X-ray methods (pict. 72)

The fronto-occipital view shows a skull with moderately large, somewhat asymmetric orbits. The right orbit is more rectangular, the left more trapezoid. The nasal root is moderately broad, and the nasal septum, which is straight, is fully preserved. Unexpressed Pacchionian bodies are present. The mastoids differ somewhat, the left being larger than the right.

On the lateral views one can see, in the course of the coronal suture, the pronounced grooves of the diploic veins. The skull is thicker occipitally than frontally. The mastoids are symmetrical and without particularities.

The pictures of the base show the expressed groove of the sigmoid sinus on the right.

All views reveal complete obliteration of all cranial sutures.

No other X-ray particularities are observed on skull no. 17.

no. 17: ♂, over 60 years

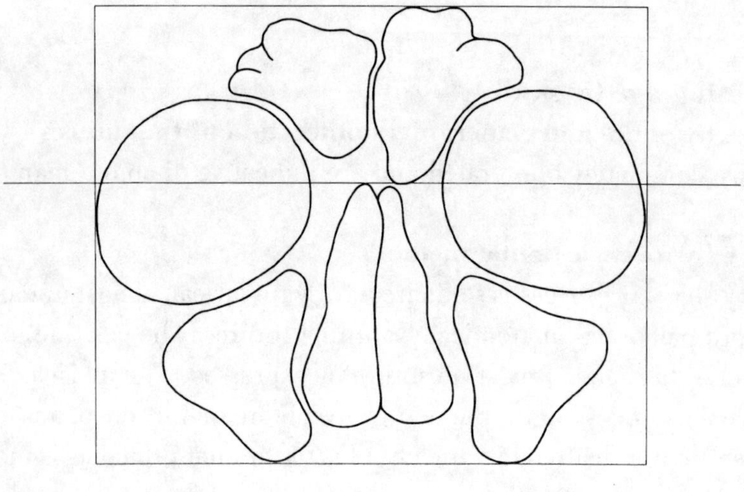

Picture 73: *Outline of the facial cavities of skull 17 from the occipito-frontal radiograph*

4.2.1.17.3. X-ray determination of the facial cavities (pict. 73)

The occipito-frontal radiograph shows that:

- the frontal sinuses are asymmetric and fan shaped;
- the orbits are somewhat asymmetric in shape, the right being more angular, the left more trapezoid. After Nemeskéri they are scored as (+1);
- the maxillary sinuses are relatively symmetrical and triangular in shape;
- the nasal cavity is symmetrical and scallop shaped with serrated margin.

Planimetric values of the facial cavities

Skull		Frontal sinuses		Orbits		Maxillary sinuses		Nasal cavity
		Right	**Left**	**Right**	**Left**	**Right**	**Left**	
No. 17	S	6.41	7.12	14.51	13.98	7.21	7.48	15.58
	H	3.65	3.65	4.19	4.30	3.56	3.74	6.32
	B	2.76	2.85	4.01	3.83	2.85	2.94	3.39
	T	0.72	0.72	0.89	0.81	0.72	0.72	0.45
	Σ	13.80						
	Sh	0.54	0.63	0.89	0.81	0.63	0.72	0.63

Legend: *S = surface, H = maximal height, B = maximal breadth, T = B/H, Σ = total surface of frontal sinuses, Sh = shape of sinuses*

4.2.1.17.4. Paleopathological methods

The left articulating surface of the temporomandibular joint is very porous and shows significant arthropathic changes. The right joint has fewer degenerative changes and its surface is smoother. Skull no. 17 does not display any other paleopathological features.

4.2.1.17.5. Commentary

According to the measurements and to the cranial index, this skull is of medium length. It is more rounded in shape, moderately heavy and little damaged. The cranial features which determine gender are less convincing than in the robust skulls 14 or 15, nevertheless they do to a great extent conform to the criteria for a somewhat less expressed male head. This conclusion is supported by the features on the frontal part: glabella, supraorbital arches, frontal tuber, and orbital rim. The occipital bone features also indicate a moderately developed occipital protuberance and occipital relief, and the two mastoid processes are of particularly masculine form. Likewise the rounded occipital opening is indicative of the male sex. Beside it to the left dorsally is an exostosis, and by the left occipital condyle there is a bean-sized hollow. The occipital condyles are symmetrical.

The facial part is determined by: the expressed rectangular or trapezoid orbits, which according to the orbital indices are characteristic of males, the moderately developed nasal root with a straight frontal part of the nasal septum (the back part is broken off), and proportionate cheekbones (the right is missing the temporal process, and both lack the zygomatic arch). Campers' angle indicates a slightly protuberant maxilla, and the dental index after Pont a semicircular palatine process of the maxilla. The dental alveoli are partially ossified, since the person had lost several teeth during lifetime. The above features indicate a symmetrically structured face of medium breadth.

There are no pathological particularities on this skull. The cranial sutures are throughout obliterated; from the radiographs it can be seen that the bone is thicker occipitally than frontally, and also that the skull is porous and its structure thinned, all of which is indicative of an older man.

THE SKULL BELONGED TO A MAN, AGED OVER 60 YEARS

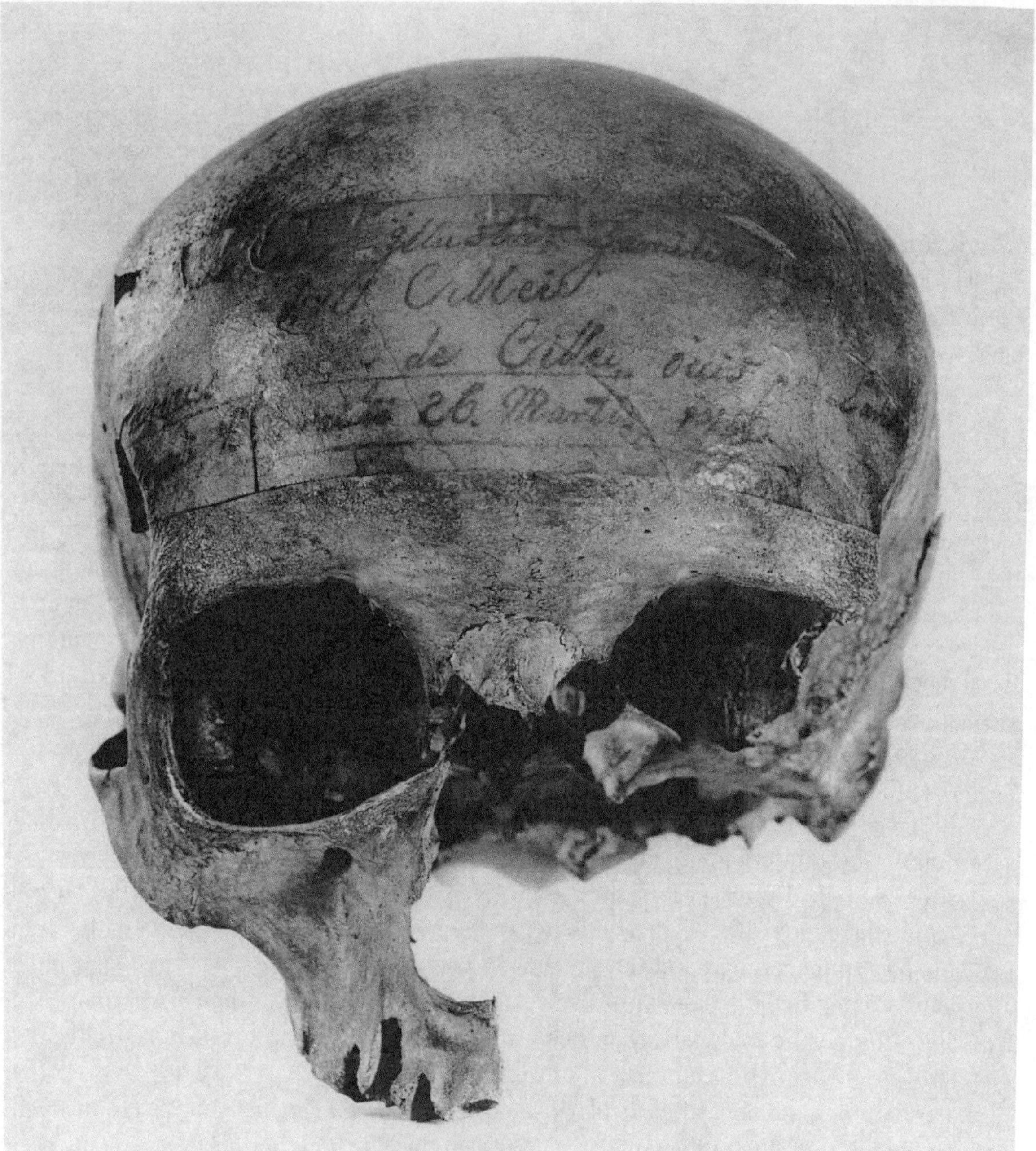

Picture 74: *Skull 18, frontal view. The skull is missing the left side of the face, supposedly damaged by a sword blow. On the forehead it bears the inscription: Ultimus ex illustri familia Comitum de Cillie occisus per Ladislavum Hunyadi Belgradii 26. Martii 1456. (Last of the illustrious family of the Counts of Celje, whose murder was ordered by Ladislaus Hunyadi in Belgrade, on 26th March 1456.)*

4.2.1.18.

Skull 18 (picts. 74, 77)

4.2.1.18.1. Determining gender by skull

Anthropological characters for determining gender

superciliary arches	moderately developed (0)
glabella	moderate (0)
frontal tubers	strongly expressed (+2)
inclination of the forehead	moderately steep (+1)
supraorbital margin	moderately expressed (+1)
orbital shape	rectangular (+1)
occipital relief	moderately pronounced (+1)
ext. occipital protuberance	considerably expressed (+1)
mastoid processes	r. stronger, l. partially damaged (+2)
cheekbones	r. medium developed (0), l. broken off
maxillary palatine process	l. half missing, r. more ellipsoid in shape
maxillary alveolar process	l. missing, r. narrower
thickness of calvaria	occipitally greater than frontally
facial part of skull	considerable part missing, preserved part seems moderately
	developed

Determining gender by anthroposcopic and anthropometric methods and indices

bizygomatic breadth	? (injury)
great occipital foramen	length 33 mm, breadth 32 mm
mastoids	both preserved, breadth r. 26 mm,
	l. has a cut off /?/ tip 15 mm
breadth of anterior nasal aperture	24 mm (after reconstruction)
cranial length	180 mm
cranial breadth	141 mm
cranial base	around 102 mm
cranial index	78.33 (medium length skull)
orbital height	r. 37 mm, l. broken off
orbital breadth	r. 42 mm, l. broken off
orbital index	r. 88.10, l. partially broken off,
	and the index is indeterminate

Determining gender by dental methods

Not performable, on account of the missing left maxilla

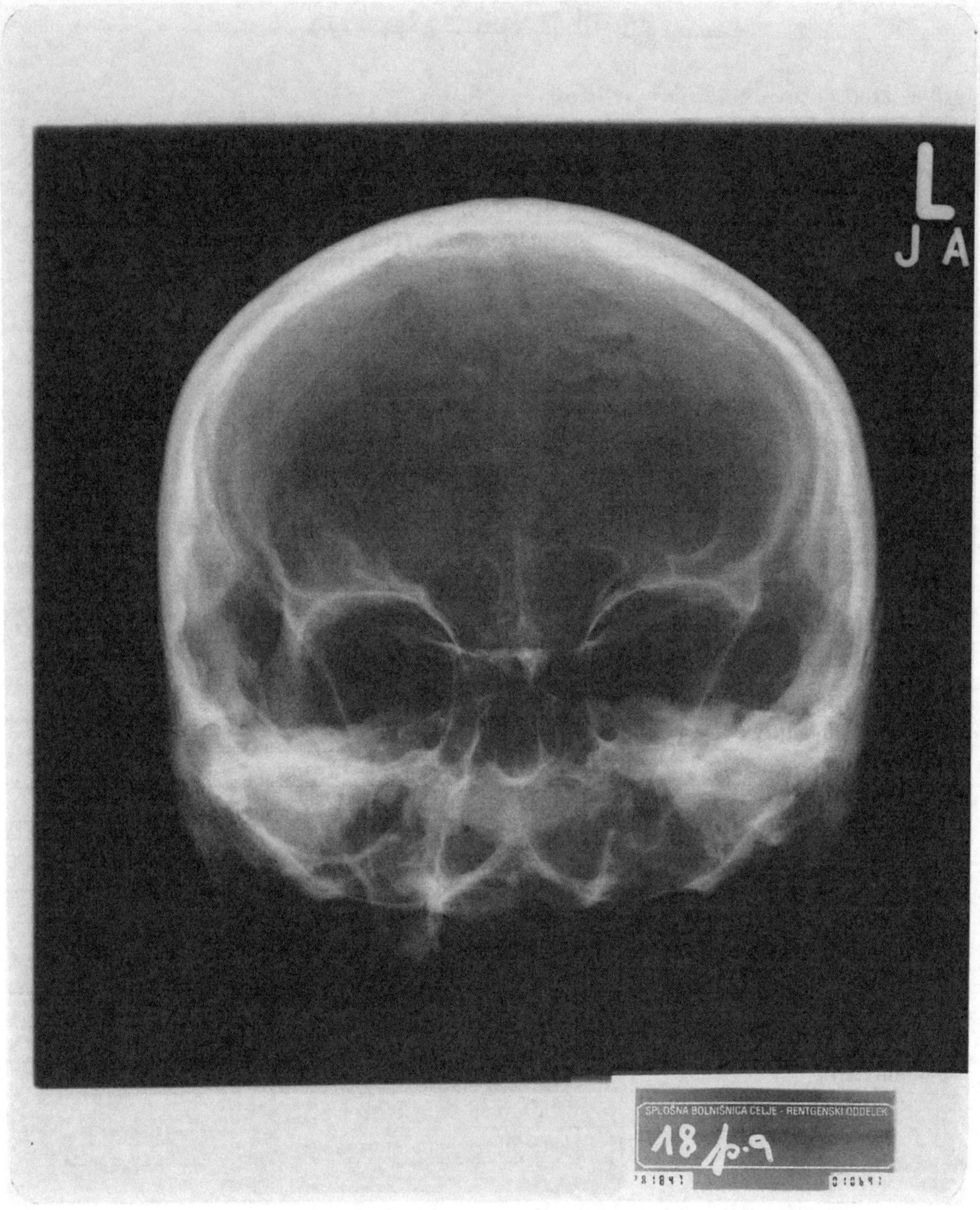

Picture 75: *Occipito-frontal radiograph of skull 18*

4.2.1.18.2. Determining age by skull

Determining age by endo- and ectocranial obliteration of the sutures

The sagittal suture is mainly obliterated. The coronal suture is still open in the distal part. The central part of the lambdoidal suture is obliterated, while distally it still remains open. The metopic suture is present. The above features indicate a person aged around 50 years.

Determining age by forensic-dental methods

Only the right maxilla is preserved; the left is broken off. There are no teeth, but all 8 alveoli are preserved.

Maxillary alveolar process from right towards left:

	M3	M2	M1	PM2	PM1	C	I2	I1	I1	I2	C	PM1	PM2	M1	M2	M3
Maxilla:	+	+	+	+	+	+	+	+	b	b	b	b	b	b	b	b

Legend: *b = broken off*

Determining age by X-ray methods (pict. 75)

The fronto-occipital views show a skull which has a relatively thick frontal bone on the vertex. The right orbit is rectangular, and the left is missing the lateral and lower walls. The nasal root is broad; the nasal septum and the nasal and ethmoid bones are broken off. Frontally visible are Pacchionian bodies and the remnants of the sagittal suture.

The side views reveal that the calvaria is thicker occipitally than frontally. Parietotemporally, the remnants of the coronal suture are visible. The bone defects earlier described are also apparent.

The images of the cranial base also reveal the bone defects of the left half of the face.

no. 18: ♂, around 50 years

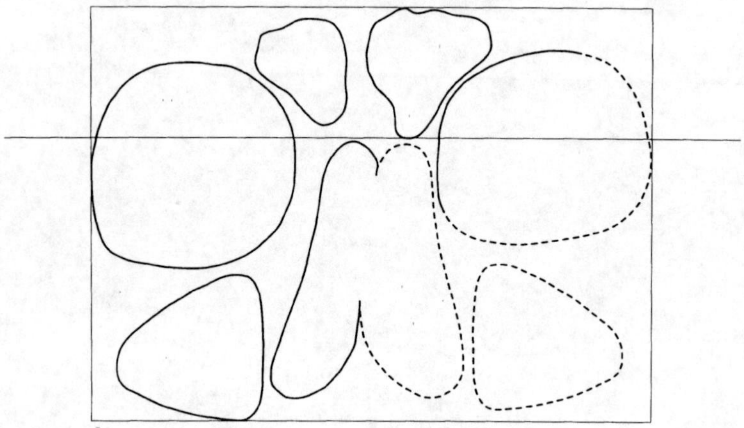

Picture 76: *Outline of the facial cavities of skull 18 from the occipito-frontal radiograph (dotted lines indicate the missing part)*

4.2.1.18.3. X-ray determination of the facial cavities (pict. 76)

The occipito-frontal radiograph shows that:

- the frontal sinuses are relatively small, asymmetric, the right is smaller than the left, they are expressed and fan shaped;
- the orbits are rectangular. After Nemeskéri they are scored as (+1);
- the maxillary sinuses are relatively symmetrical. They resemble the shape of a triangle;
- the nasal cavity is symmetrical; its shape is more difficult to determine, since the skull is severely damaged.

Planimetric values of the facial cavities

Skull		Frontal sinuses		Orbits		Maxillary sinuses		Nasal cavity
		Right	Left	Right	Left	Right	Left	
No. 18	S	3.20	4.90	14.00	14.30	7.30	6.30	14.90
	H	2.30	2.90	4.70	4.70	3.50	3.50	6.30
	B	2.00	2.40	4.00	4.10	3.00	2.60	3.90
	T	0.80	0.80	0.90	0.90	0.90	0.70	0.60
	Σ	8.10						
	Sh	0.90	0.80	0.90	1.00	0.90	0.80	0.50

Legend: *S = surface, H = maximal height, B = maximal breadth, T = B/H, Σ = total surface of frontal sinuses, S = shape of sinuses*

4.2.1.18.4. Paleopathological methods

On skull 18 the most striking are the changes purportedly caused by a cold weapon. According to the historical records, this is supposedly the skull of the last male descendant of the Counts of Celje, Ulrich II of Celje, who was murdered in 1456 in Belgrade. The injury incurred belongs to the domain of forensic medical examination.

The articulating surfaces of temporomandibular joints left and right are without the typical arthrotic changes.

4.2.1.18.5. Forensic medical expertise

The skull is massive, dark coloured, lacquer coated. On the forehead is affixed a deeply yellowed inscription in black ink with a text in Latin (pict. 74).

The historical records relate that the last male descendant of the Counts of Celje was killed in Belgrade in the year 1456, and that when already dead he was decapitated by sword. The conviction that skull 18 belonged to Ulrich II of Celje prevails. According to historiographic data, he was aged 50 at the time of his death, and his corpse was said to have been returned to Celje, where he was buried in the crypt of the Counts of Celje in St Mary's church.

Skull no. 18 has a huge injury to the facial part of the bone on the left side. The mandible apart, it is also missing the whole of the left maxilla, the left half of the hard palate, the upper wall of the left orbit, all of the left cheekbone, half of the left mastoid, and the left ethmoid. The maxillary frontal process is also broken off on the right. The tip of the left mastoid is cut off in such a way that the surface is quite straight and allows conjecturing regarding the direction of the blow of the blade, which runs at an angle of approx. 45 degrees in relation to the sagittal axis of the body. The injury was caused by a long blade, which could have cut through the entire left lateral part of the skull. Considering both the legend and the expertise conducted, this weapon could have been a sword. The injury was presumably fatal.

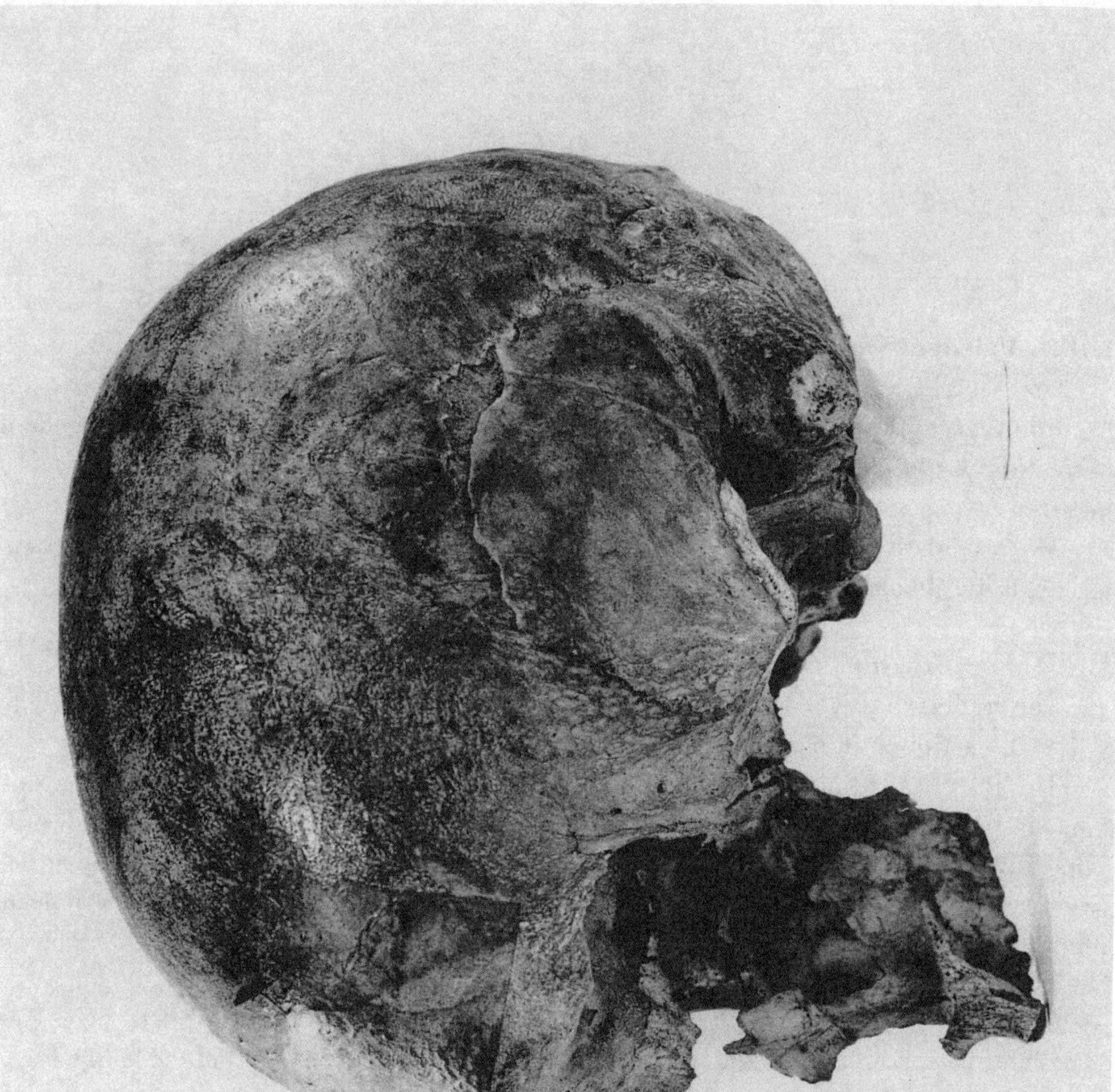

Picture 77: *Skull 18, side view*

4.2.1.18.6. Commentary

According to the cranial index, this skull is of medium length. It is moderately heavy, dark coloured, and lacquer coated (probably for protection).

The skull is severely damaged, presumably by a sword blow. According to forensic medical expertise the consequences were fatal.

It may be recorded that, by its structure, this is a relatively gracile male skull. The features of its frontal part are: poorly expressed glabella and supraorbital arches, and better expressed frontal tuber on a relatively high forehead. The occipital part of the skull is robust, as is evidenced by the mastoids and the occipital relief with the protuberance. The great occipital foramen is more rounded in shape. The condyles of the occipital bone are asymmetric: the left is smooth, the right tilted by approx. 5 mm to the right. The person probably held his head to one side (tilted to the right).

The right orbit is rectangular, the orbital index is of masculine type. The nasal root is moderately wide, the septum is not preserved. At the insertion of the mastoid muscle left, the maxillary zygomatic process is broken off (25 mm x 4 mm). As a consequence, the entire left maxilla with the palatine and frontal processes and the ethmoid bone were broken off. Campers' angle indicates the orthognathic position of the maxilla. The preserved part of the maxillary palatine process indicates a rather narrow jaw which, together with narrowish cheekbones, creates a narrow face. On the right side of the dental arch all the dental alveoli are preserved, indicating that the man had healthy teeth and that they fell out posthumously.

The cranial sutures are mainly obliterated. From the centre of the forehead upwards runs a suture, the so-called metopic suture, which is rarely seen, and is a remnant of the embryonic origin of two parts of the frontal bone. The articular surfaces of the maxillary joint are without particularities.

The above described features indicate the gracile skull of an adult male, aged about 50.

THE SKULL BELONGED TO AN ADULT MALE, AGED ABOUT 50 YEARS

4.2.2.
Final review of gender and age of the studied skulls 1–18

skull	gender	age
1	female	over 50 years
2	child	around 10 years
3	female	30–40 years
4	female	over 50 years
5	male	around 50 years
6	male	over 50 years
7	male	30–40 years
8	male	around 30 years
9	male	30–40 years
10	female	30–40 years
11	female	around 40 years
12	female	around 30 years
13	male	around 40 years
14	male	around 40 years
15	male	30–40 years
16	child	around 10 years
17	male	over 60 years
18	male	around 50 years

4.3.
Comparison of the Historical and Identificational Results

4.3.1.
Principle for Numbering of the Skulls

All the skulls are numbered from 1 to 18. The last skull is missing half of the face, and according to historical sources it belonged to Ulrich II of Celje. Who numbered the skulls, and when, is not known. It is possible that this was done after the second fire in St Mary's church, when the skulls were transferred by the schoolteacher of Latin, Anton Zupančič, from the former vault behind the altar. It is also possible that he wrote the inscription in Latin and affixed it to the forehead of skull 18.

One has to constantly keep in mind the awareness that in feudal society the hierarchical order prevailed, and that it was respected. It is therefore difficult to conceive that it might have been any different with the skulls of the rulers. Nevertheless, one cannot avoid the conjecture that the skulls might also have been numbered without proper order.

Even more remote is the thought that the skulls studied might possibly have been mixed with others from the remaining two crypts in St Mary's church – the minorite and the crypt of the arms-bearers of the Counts of Celje, the Knights of Soteška. In fact, in the church there are no skulls preserved, either of the former or of the latter.

The results of the identification study indicate that the order of interment followed the time sequence of dying, with separate burial of male and female members of the family.

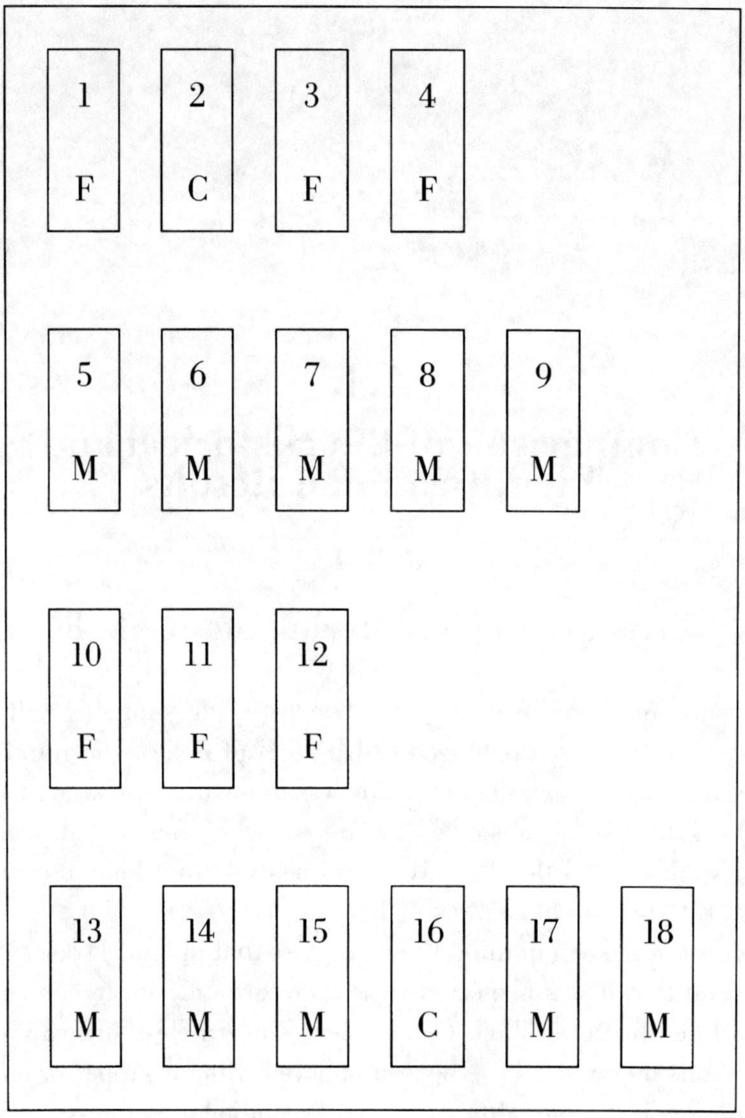

Picture 78: *Hypothetical appearance of the Celeian vault. It was around 15 m long, approx. 10 m wide. The numbers are identical with those on the skulls*
Legend: *F – female, M – male, C – child*

4.3.2.
Vault of the Counts of Celje

Concerning the crypt, the sources record only that it was beneath the presbytery of the church. Today it is no longer preserved. It is not known when it was emptied. Probably after the second fire in the church it happened for the first time. It is not known whether they left the lead coffins inside. They must have been later melted down by the soldiers for military purposes. According to eye-witness accounts, the vault was approximately 10 x 15 metres in size, and completely empty. It had been destroyed following an explosion of gas in around 1970.

Hypothetical appearance of the Celeian family vault (pict. 78):
The coffins follow in order from left to right.

Skulls by gender in the first row:
1 – female (f), 2 – child (c), 3 – f., 4 – f.;

Skulls by gender in the second row:
5 – male (m), 6 – m., 7 – m., 8 – m., 9 – m.;

Skulls by gender in the third row:
10 – f., 11 – f., 12 – f.;

Skulls by gender in the fourth row:
13 – m., 14 – m., 15 – m., 16 – c., 17 – m., 18 – m.

The identification findings reveal that the first four skulls are of female gender (1–4), and the second among them is a child's skull; that the next five skulls (5–9) are male; they are then followed by three female skulls (10–12), and another six male skulls (13–18), including one of a child. From this it may be concluded that in the vault the males were buried separately from the women and children. It may also be possible that the children of the dynasty were interred with the male members of the family, and the girls with the female members.

From the technical point of view, in the vault, sized 10 x 15 m, it would certainly have been possible to place four rows of coffins along its length. If one assumes that the coffins were 2 m long and 1 m wide, then longitudinally, with one-metre intervals, they would have occupied approximately 15 metres. The same in width – if the space was approximately 10 m wide, and if each coffin was approx. one metre wide – such an arrangement would have been suitable.

Considering the identification results, with comparison of the facial cavities and other epigenetic traits, it may be supposed that the skulls must have been numbered upon removal from the vault, and that their ordering is a chronological indication of the interment of the counts, only that in this procedure the dynasts were separated from their wives and children.

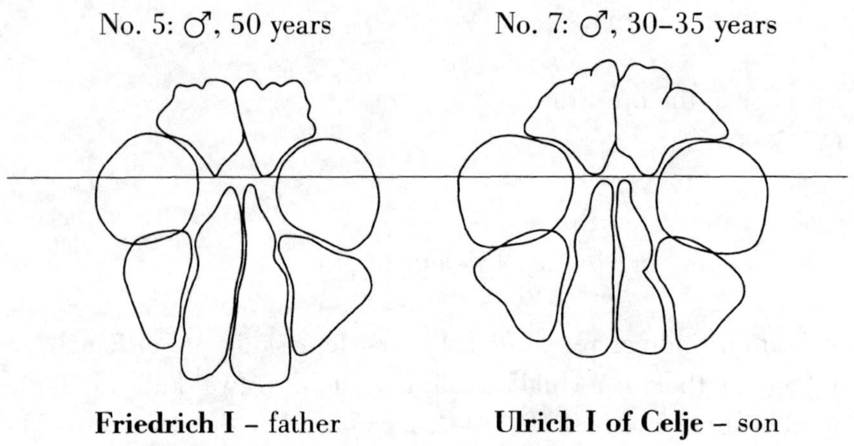

Picture 79: *Facial cavities of skulls 5, 7, 6 and 1*

4.4.
Results on Kinship

Epigenetic identification by skulls was carried out: the shape and size of the facial cavities and around 30 other epigenetic (»discrete«) skull traits were compared. These results were then compared with conclusions from the historical and genealogical data. Although the natural science findings provided substantial results, additional conclusions had to be drawn because of the lack of required reliable historical sources. It was only by comparing the natural science, historiographic and genealogical data, and by following the principle of numbering of the skulls, that the results could be achieved. Since it was supposed that the skulls belonged to blood relatives, and since the principle behind their numbering had been hypothetically resolved, the author tried to verify these hypotheses by the appropriate methods.

4.4.1.
The Family of Friedrich I (skulls 5,7,1 and 6)

It was expected that among the skulls would be found that of Friedrich I, the founder of the vault, and of his closest kin the skulls of his sons Ulrich I of Celje and Hermann I of Celje. It was also anticipated that the skull of Friedrich's wife, Didmund of Walsee, would be included. Since it was our opinion that the family members were interred in temporal order in relation to the occurrence of death – although with separate burial of the male and female members – it was foreseen that the first male skull (5) would be that of Friedrich I (died 1359 or 1360). Following the order of interment, among the males Friedrich I was followed by his son Ulrich I of Celje (died 1368), therefore it was expected that skull 6 would belong to him. The second son, Hermann I of Celje, would supposedly have been buried as the following adult male (in 1385), therefore it was expected that his skull would be no. 7. Identification by age revealed skull 6 is aged over 50 (up to 60) years, while skull 7 was assessed at an age of 30–40 years. It is known from the historical data that Ulrich I of Celje died young, at the age of 30, while Hermann I of Celje lived on to see mature age. Hence, it was to be expected that these two skulls would be numbered in reverse, as was indeed confirmed by further identification.

Friedrich's wife Didmund was most probably the first to be interred in the vault. According to the sources, she died either in 1353 or in 1354. It was therefore expected that the first female skull (1) belongs to her.

From the postero-anterior images of skulls 5, 7, 1 and 6, outlines of the facial cavities were drawn, and crosscomparisons were made regarding shape and size (pict. 79). Since this epigenetic identification involved illustration of the facial cavities and other epigenetic traits, the first family was shown in closer detail than the others.

The frontal sinuses of skulls 5 and 7, are very similar in their fan shape and size as the rectangular orbits are. The maxillary sinuses are triangular in shape in both cases and have comparable surfaces. The shape of the nasal cavities in both skulls is similar to a scallop shape with serrated margins, while their surface in skull 5 is somewhat larger than in skull 7.

Since skulls 5 and 7 are very similar concerning the shape and size of the facial cavities, it may be concluded that these are traits of an inherited relation: it is highly probable that they belong to the father and son, Friedrich I (5) and Ulrich I of Celje (7).

Kinship between skulls 5, 7, 1 and 6 was further verified.

The facial cavities of skulls 5 and 6 were compared. They have similar rectangular orbits and scallop shaped nasal cavities, and comparable triangular maxillary sinuses. In skull 6, the orbits, maxillary sinuses and nasal cavities are larger than in skull 5. Only the frontal sinuses differ more notably in size: both are, otherwise, fan shaped, although in skull 5 they are more wide-spread. It may be concluded that the two skulls compared are, with regard to the characters of the facial cavities, inter-related by inheritance. They supposedly belong to the father and son, Friedrich I (5) and Hermann I of Celje (6).

The male skulls 5, 6 and 7 are similar in shape and size of the facial cavities: all three have rectangular orbits, triangular maxillary sinuses, and scallop shaped nasal cavities. If one compares the surfaces of the three facial cavities, they are also similar. Only the frontal sinuses of skull 6 are not prominently fan shaped, and their surface is somewhat smaller. From the above it may be concluded that the skulls are interrelated by kinship, and that they belong to the father and the two sons, Friedrich I, Hermann I of Celje and Ulrich I of Celje.

The shape and size of the facial cavities of skulls 1 and 6 were also compared. Both have highly similar, but smaller frontal sinuses. In both, they are fan shaped. Since skull 1 is female and skull 6 male, it follows from the theoretical presuppositions regarding the size of all facial cavities that they are smaller in women. The orbits of skull 6 are more rectangular in shape and more similar to those of skull 5 than of skull 1, in which they are more rounded. The nasal cavity in skull 6 is, likewise, more similar to that of skull 5, for in both cases it resembles a scallop with serrated margin. The female skull no. 1 has a pear shaped nasal cavity. In all four skulls the maxillary sinuses are triangular.

Skulls 1 and 6 are highly similar, particularly in the shape of the frontal sinuses. They are also alike in the size of the facial cavities. Skull 6, in comparison with skulls 5 and 7, has a smaller paranasal cavity surface.

The above-described features of the facial cavities of skulls 5, 7, 1 and 6 indicate that they are genetically interlinked. Skull 5 belonged to Friedrich I, skull 1 to Didmund of Walsee, skull 7 to Ulrich I of Celje, and skull 6 to Hermann I of Celje. Among them they were the father (5), the mother (1) and the two sons. The son Ulrich I of Celje (7) inherited the shape and the size of the facial cavities mainly from his father (5), while the son Hermann I of Celje (6) inherited particularly the shape of the frontal sinuses and the size of the facial cavities from his mother (1).

It may therefore be concluded that the son Hermann I of Celje inherited the characteristic facial cavities more from his mother, while the son Ulrich I of Celje inherited them more from his father.

Skulls 5, 6, 7 and 1 were compared also by their shaping features (pict. 80):

According to the cranial index, these male skulls are all on the borderline between short and medium length. The female skull is short.

- the three male skulls are rounded in shape, the female one is pentagonal;
- skulls 5, 1 and 6 are of medium facial breadth, skull 7 is broader;

– all three male skulls have a broad nasal root;

– they all have the great occipital foramen alike, i.e. more rounded;

– the shape of the temporal bones in the male skulls 5, 6 and 7, towards the skull vertex, is highly semicircular with a conical peak (see picts. 5, 6, 7);

– it is known that dysgnathia is inherited (e.g. the Habsburgs). It may be recorded that the three male skulls are orthognathic, while the female skull is slightly prognathic, with a Campers' angle of 80 degrees.

Survey of the most important epigenetic traits of skulls 1, 5, 6 and 7

- supraorbital foramina and supratrochlear notches (pict. 84)
  - 1 – right notch, left notch
  - 5 – right foramen, left notch
  - 6 – right notch, left notch
  - 7 – right notch, left notch
- (double) infraorbital foramina (pict. 85)
  - 1 – left and right double
  - 5 – left and right single
  - 6 – left and right single
  - 7 – left and right single
- double zygomatic-facial foramina (pict. 15)
  - 1 – right 4 foramina, left 1
  - 5 – right and left double
  - 6 – right and left single
  - 7 – right and left double
- parietal foramina (pict. 96)
  - 1 – right 1
  - 5 – left 1, right 1
  - 6 – left 1, right 1
  - 7 – right 2
- extrasutural mastoid foramina
  - 1 – left 1
  - 5 – left 2; right 1
  - 6 – right 1
  - 7 – left 1; right 1
- supranasal sutures (pict. 97)
  - 1 – right 1, left 1
  - 5 – right 1, left 1
  - 6 – right 0, left 1
  - 7 – right 1, left 1
- transverse palatine suture and torus palatinus (pict. 86)
- marginal tubercle (thickening of the temporal part of the frontal part of the cheekbones) is expressed:
  - 1 – right
  - 5 – not present
  - 6 – right and left
  - 7 – right
- pharingeal tubercule
  - 1 – slightly expressed
  - 5 – moderately expressed
  - 6 – slightly expressed
  - 7 – not present

Above are presented the central epigenetic traits which can be observed on skulls 1, 5, 6 and 7.

Interpretation: the supraorbital features are mutually almost identical. The infraorbital foramina are identical in skulls 5, 6 and 7; only skull 1 has a double foramen. In the zygomatic-facial foramina skulls 5 and 7 have the same features, while skulls 1 and 6 are similar.

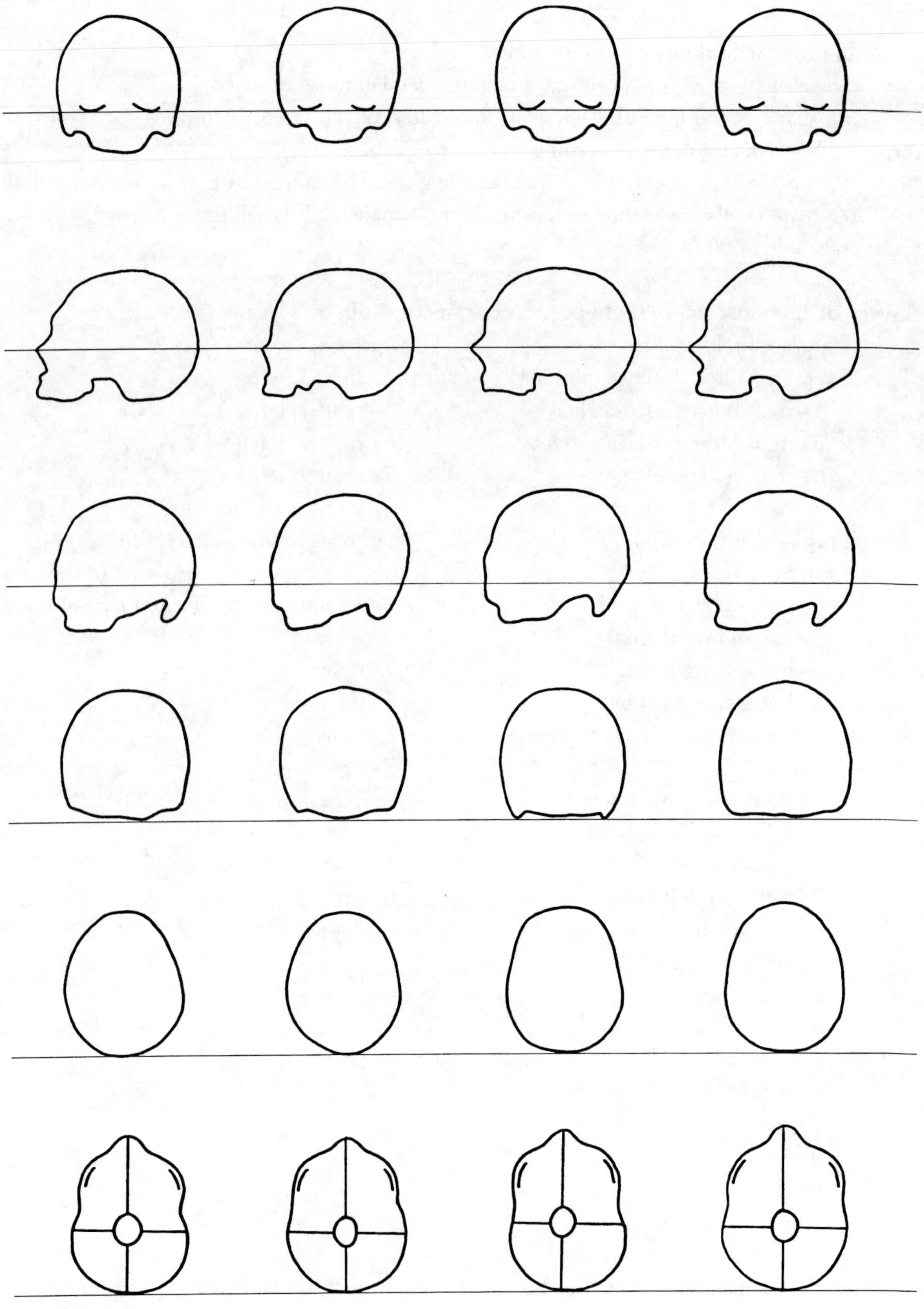

Picture 80: *Outline of skulls 5, 1, 6 and 7 from the front, from the side, in semi-profile, from behind, from above, and from below (sketched from the photographs by V. Berk)*

The parietal foramina are of equal number in skulls 5 and 6; in skulls 1 and 7 there are, respectively, one and two on the right. According to the literature these are inherited in an autosomal dominant manner. The extrasutural mastoid foramina do not indicate any greater variability. In all four skulls, the supranasal sutures are almost identical, while the palatine transverse suture is more similar between skulls 6 and 7, and also does not essentially differ in skull 5. In skull 1, the palatine transverse suture is somewhat different. The marginal tubercle may be seen in skulls 1, 6 and 7, but is not present in 5. The faringeal tuberculum may be traced in skulls 1, 5 and 6, but not in 7.

The temporal bones are also features worth mentioning. In the male skulls 5, 6, and 7 they are cupola shaped with a pointed peak (picts. 33, 36, 39). In this respect there is more similarity between the temporal bones of the male skulls 5 and 7, and between the female skull 1 and the male 6. Also in this case the son Hermann I of Celje (6) inherited this trait from his father, while the son Ulrich I of Celje (7) inherited it from his mother (1).

The kinship between the skulls was also confirmed according to the criteria of differentiation of the skull sutures: the basal configuration is easily comparable, particularly on the coronal suture. In the male skulls 5, 6 and 7 it is serrated (dentata). Likewise the nasal shapes are suitable for comparison: skulls 1 and 5 have a straight nasal bone, while in 6 and 7 the shape is straight to concave.

The similarity between the skulls, as revealed by the facial cavities, is further substantiated by taking into account the epigenetic traits. It can be observed that, in the great majority of cases, these characters between skulls correspond to and confirm the kinship. This is especially true for the three male skulls studied.

Considering the identification, historiography, genealogy and numbering of the skulls:

SKULLS 5, 1, 6 AND 7 BELONG TO ONE FAMILY, WITH A HIGH DEGREE OF PROBABILITY TO THE FATHER FRIEDRICH I (5), HIS WIFE DIDMUND (1), AND THE TWO SONS HERMANN I OF CELJE (6) AND ULRICH I OF CELJE (7).

4.4.2.
The Family of Count Hermann I of Celje (skulls 6, 4 and 2)

Following the already presented principle of epigenetic identification, the family linkage between the skulls was first established.

Hermann I (6) was married to Catherine Kotromanić; they had a son, Hans, who died as a child. According to the order of interment, the first child's skull bears the number 2. Since it is not known when Catherine Kotromanić died, it would not have been possible to predict from the historical sources which skull would be hers. By comparing the shapes and sizes of the facial cavities and the epigenetic traits, the following conclusions were reached:

From the occipito-frontal radiographs of skulls 6, 4 and 2, sketches were made of the outline of the facial cavities, and these were compared by shape and size (pict. 81):

The fan shaped frontal sinuses with the undulating margins of female skull 4 are similar in shape to the same frontal sinuses of child's skull 2. Likewise, the male frontal sinuses are fan shaped, only their margins are smoother. The orbits of the female and the male skull are also similar to those of the child's skull, more rounded, and in the child's skull smaller in surface. The maxillary sinuses in the male and female skulls are triangular in shape, while in the child's skull they are still in the developmental phase, as is the case for the other features of the facial cavities. The shape of the nasal cavity in male and female skulls resembles a scallop with a serrated margin; the child's nasal cavities are still not fully shaped, although they are already close to being scallop shaped.

Comparison of the facial cavities confirmed the hypothesis that child's skull 2 bears the features of male skull 6, that of the hypothetical father, and female skull 4, the hypothetical mother.

Other epigenetic traits for skulls 6, 4 and 2

- supraorbital foramina and supratrochlear notches (pict. 84)
  - 6 – right notch and foramen, left notch
  - 4 – right notch, left notch
  - 2 – right notch, left foramen
- (double) infraorbital foramina (pict. 85)
  - 6 – left and right single
  - 4 – left and right single
  - 2 – left and right damaged
- double zygomatic-facial foramina
  - 6 – right and left single
  - 4 – right and left single
  - 2 – right injury, left double
- parietal foramina (pict. 96)
  - 6 – right 1, left 1
  - 4 – right 1
  - 2 – right 1, left 1
- extrasutural mastoid foramina
  - 6 – right 1
  - 4 – 0
  - 2 – left 1, right 1
- transverse palatine suture and palatine tori (pict. 86)
- expressed marginal tubercle
  - 6 – right and left
  - 4 – left
  - 2 – slight on the left

No. 6: ♂, 50–60 years | No. 4: ♀, 50–60 years | No. 2: ♂, 12 years

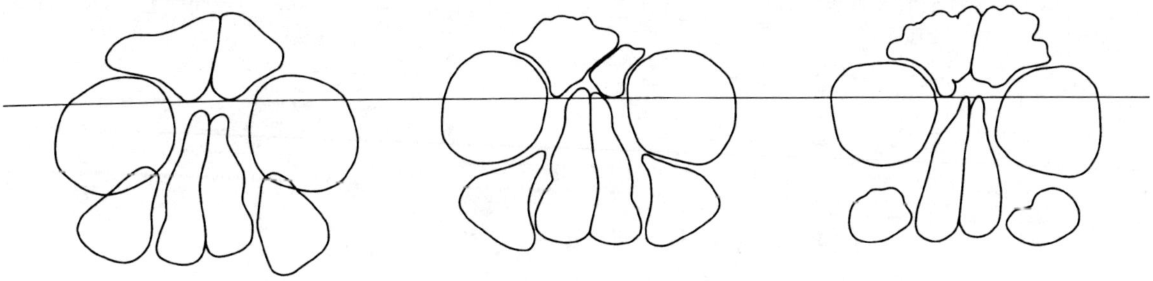

Hermann I of Celje – father | **Catherine Kotromanić** – mother | **Hans** – son

Picture 81: *Facial cavities of skulls 6, 4 and 2*

Skull 2, has next to the so-called lambda (the juncture between the sagittal and lambdoid sutures), two independent lambdoid ossicles. Likewise, in the course of the lambdoid suture left and right, it has one small sutural ossicle on each side (picts. 19, 24).

Since additional ossicles on the skull are rare, and are proven to be dominantly hereditary with a penetration of approx. 50%, they constitute an important epigenetic trait. Consequently, an examination was made to determine in which skulls they were still present:

skull 2: 2 lambdoid ossicles and sutural lambdoid ossicles, 1 left and 1 right,
skull 9: ossicle right and left in the lambdoid suture,
skull 10: two larger lambdoid ossicles, skull 12: small ossicle left in the lambdoid suture,
skull 18: small ossicle left in the lambdoid suture.

A study was made to determine whether child skull 2 could be in kinship to any of the above-mentioned skulls. Although this could not be confirmed, the research did indicate that skulls 9 and 10 belonged, with a high degree of probability, to the son (9) and the mother (10).

Interpretation: the supraorbital features are the same for skulls 4 and 6 – a combination of notches and foramina which was also inherited by skull 2. In all three skulls the infraorbital foramina are identical. In the zygomatic-facial foramina, male skull 6 and female 4 have the same traits, while skull 2 differs. The parietal foramina are identical in the male and child's skulls, while in the female skull there is only one foramen (an autosomal dominant characteristic!). The extrasutural mastoid foramina occur in a more similar form in the male and the child's skulls. The palatine transverse suture is similar in all three skulls, and all three are also lacking palatine tori. The marginal tubercle is evident in all three skulls; its genetic determination has not yet been proven to a satisfactory degree.

The similarity between skulls 6, 4 and 2, as indicated by the facial cavities, acquires further substantial importance after taking into account the already surveyed epigenetic traits. Considering the identification, historiography, genealogy and skull numbering, it may be concluded that:

SKULLS 6, 4 AND 2 BELONG TO ONE FAMILY, WITH A HIGH DEGREE OF PROBABILITY TO THE FATHER, HERMANN I OF CELJE (6), THE MOTHER CATHERINE KOTROMANIĆ (4), AND THE JUVENILE SON HANS (2).

4.4.3.

The Family of Ulrich I of Celje (skulls 7, 3, 8)

Ulrich I of Celje was married to Adelaide of Ortenburg. They had a son who died in Vienna of an unknown cause, aged about 30.

Following the identification results, skull 7 would correspond to Ulrich I of Celje, and skull 8 to the younger Viljem. Regarding Adelaide of Ortenburg, the fact is known that she was interred in the family vault in 1391, as the second countess after Didmund of Walsee. Since the children were also interred together with the women, and since a child's skull is marked as no. 2, it may be expected that Adelaide's mortal remains were marked as no. 3.

Outlines of the facial cavity of male skull 7, female skull 3, and also male skull 8 were drawn from the occipito-frontal radiographs. The results obtained were carefully compared in terms of shape and size (pict. 82).

Commentary: male skulls 7 and 8 are at first sight similar in terms of the shape and size of the facial cavities. The large fan-shaped-pyramidal frontal sinuses of skull 7 have developed into pyramidal sinuses in skull 8. The surfaces of both are similar. The frontal sinuses of female skull 3 are, likewise, fan shaped, but much smaller. In all three skulls the shape of the orbits is more rectangular, although less in skull 3 than in skulls 7 and 8. Both the uniform triangular maxillary sinuses and also the scallop shape forms of the nasal bone confirm the hypothesis that these three skulls are linked by heredity.

Comparison of the facial cavities confirmed the hypothesis that male skull 8 bore the characteristics of male skull 7, which is that of the hypothetical father, and of female skull 3, the hypothetical mother.

No. 7: ♂, 30–35 years — No. 3: ♀, 30–40 years — No. 8: ♂, 30 years

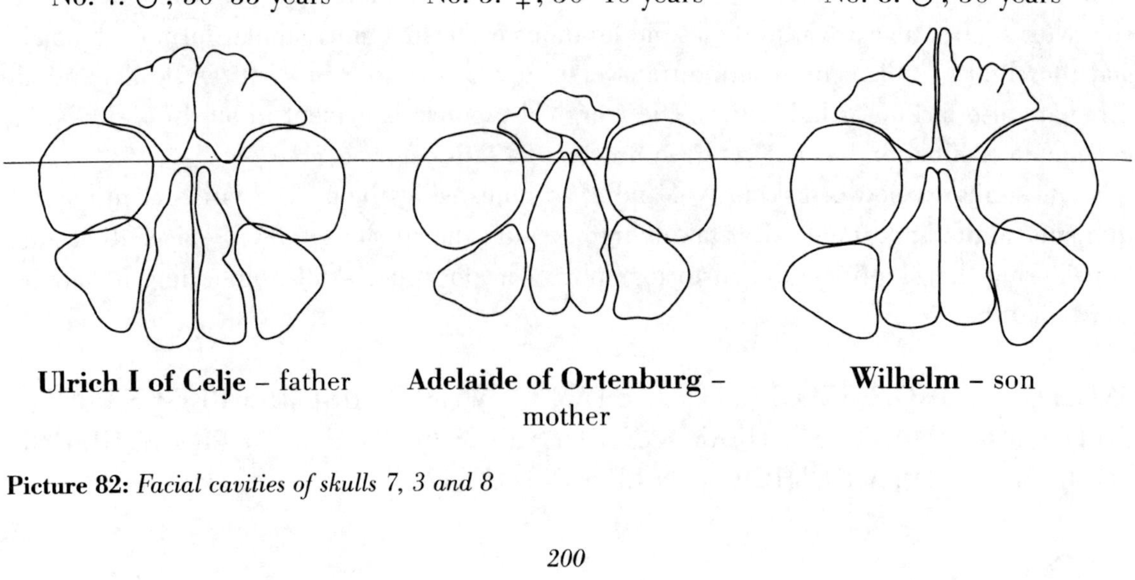

Ulrich I of Celje – father — **Adelaide of Ortenburg** – mother — **Wilhelm** – son

Picture 82: *Facial cavities of skulls 7, 3 and 8*

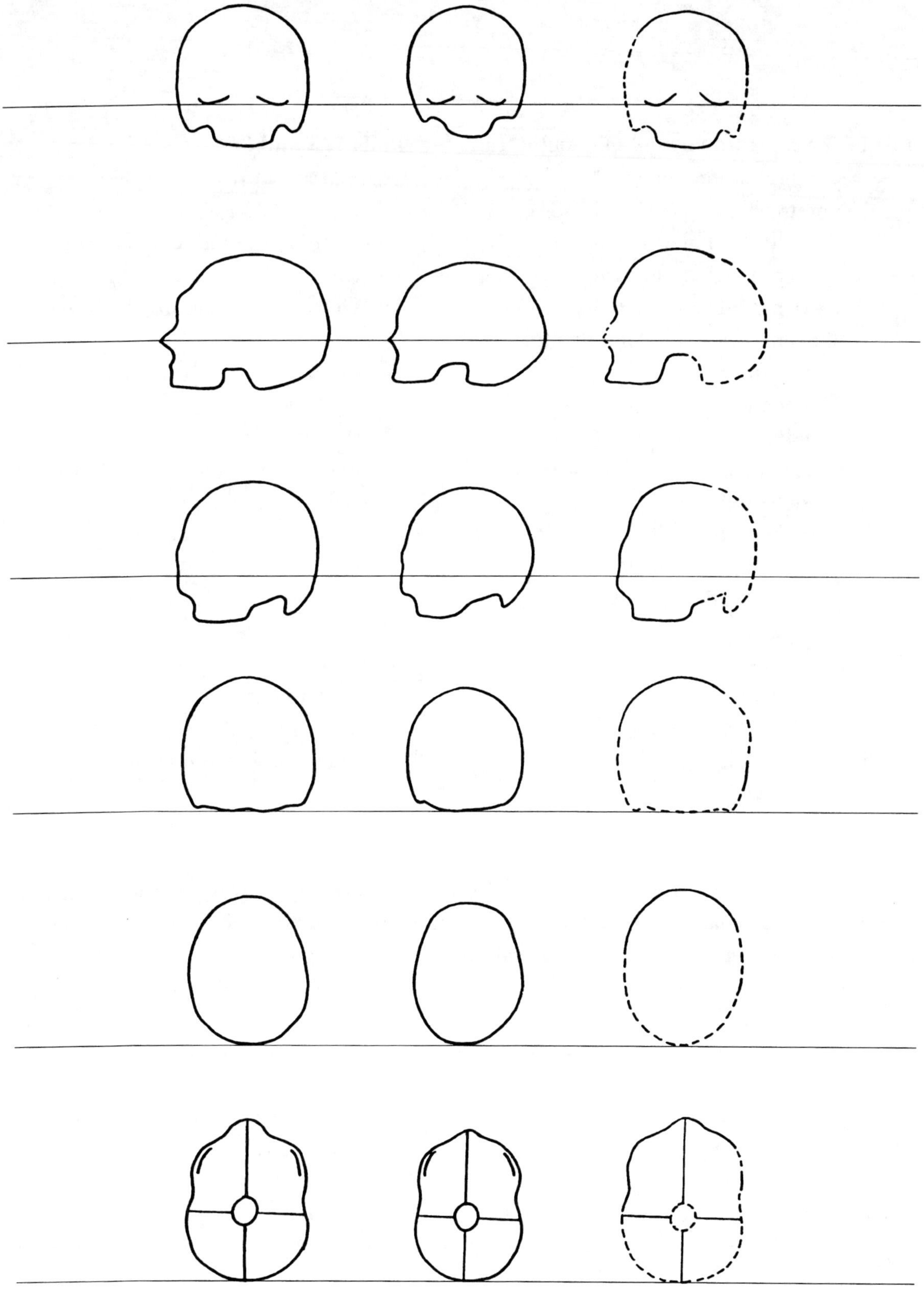

Picture 83: *Outline sketch of skulls 7, 3 and 8 from the front, the side, semi-profile, from behind, above, and below (sketched after the photographs by V. Berk)*

List of the most important epigenetic traits for skulls 7, 3 and 8

- supraorbital foramina and supratrochlear notches (pict. 84)
  - 7 – right notch, left notch
  - 3 – right notch, left notch
  - 8 – right notch, left notch
- (double) infraorbital foramina (pict. 85)
  - 7 – left and right single
  - 3 – left and right single
  - 8 – left and right double
- double zygomatic-facial foramina
  - 7 – right and left double
  - 3 – right and left single
  - 8 – right and left single
- parietal foramina (pict. 96)
  - 7 – right 2
  - 3 – right 1
  - 8 – injury
- mastoid extrasutural foramina
  - 7 – left 1; right 1
  - 3 – right 1
  - 8 – injury
- supranasal sutures (pict. 97)
  - 7 – right 1, left 1
  - 3 – small ossicle on nasal suture
  - 8 – small ossicle on nasal suture
- transverse palatine suture and palatine tori (pict. 86)
  - 3 is very similar to 8
- marginale tubercle
  - 7 – right expressed
  - 3 – not present
  - 8 – right and left expressed
- torus palatinus (pict. 86)
  - 7 – 0
  - 3 – expressed left and right
  - 8 – less expressed
- marginal process
  - 7 – moderate right
  - 3 – 0
  - 8 – expressed left and right

There are several characteristics among the epigenetic traits, which stress particularly the kinship interrelation between skulls 7, 3 and 8. Skulls 3 and 8 are the only two amongst those examined which display a rarely inherited trait, i.e. an ossicle above the nasal bone, the so-called supranasal ossicle. Furthermore, the two sutures above the nasal bone are still open in both skulls, thus indicating a younger person. This suture remains open until about the age of 30 (it is known that Wilhelm died at approximately that age). Likewise, the transverse palatine sutures in skulls 3 and 8 are extremely similar, which is also a strong sign of kinship. Skulls 3 and 8 have also palatine tori. In addition, the shaping features of the three skulls were also compared; their similarity is most eloquently testified by the accompanying picture (pict. 83). On the basis of all these signs, one may with a great measure of probability affirm that skull 3 belonged to the mother, and skull 8 to the son.

Considering the identification, historiography, genealogy and skull numbering, one may conclude that:

SKULLS 7, 3 AND 8 BELONG TO ONE FAMILY, WITH A HIGH DEGREE OF PROBABILITY TO THE FATHER ULRICH I OF CELJE (7), MOTHER ADELAIDE OF ORTENBURG (3), AND SON WILHELM (8).

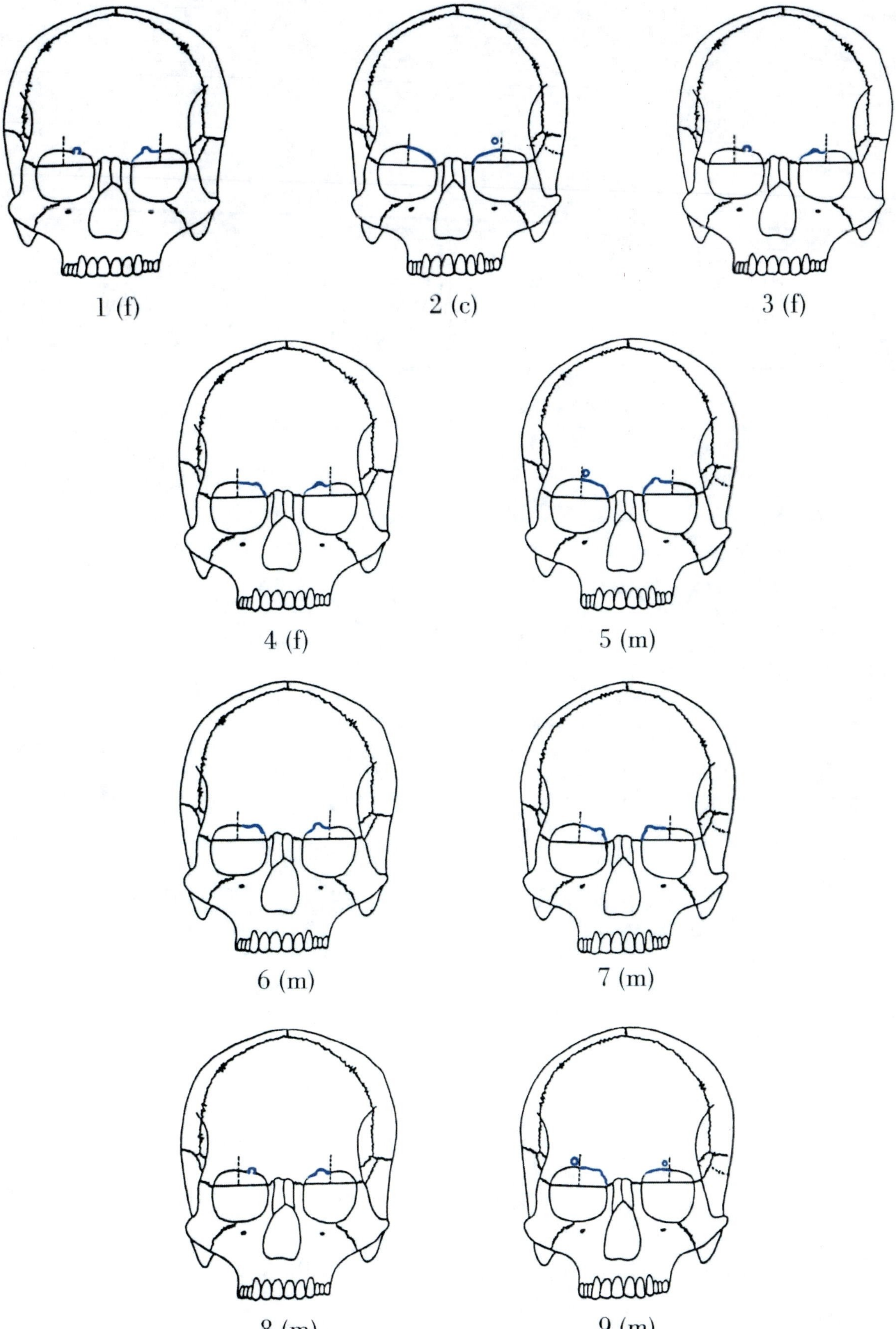

Picture 84: *Epigenetic traits – supraorbital foramina and notches of skulls 1–9*
Legend: *f = female, m = male, c = child*

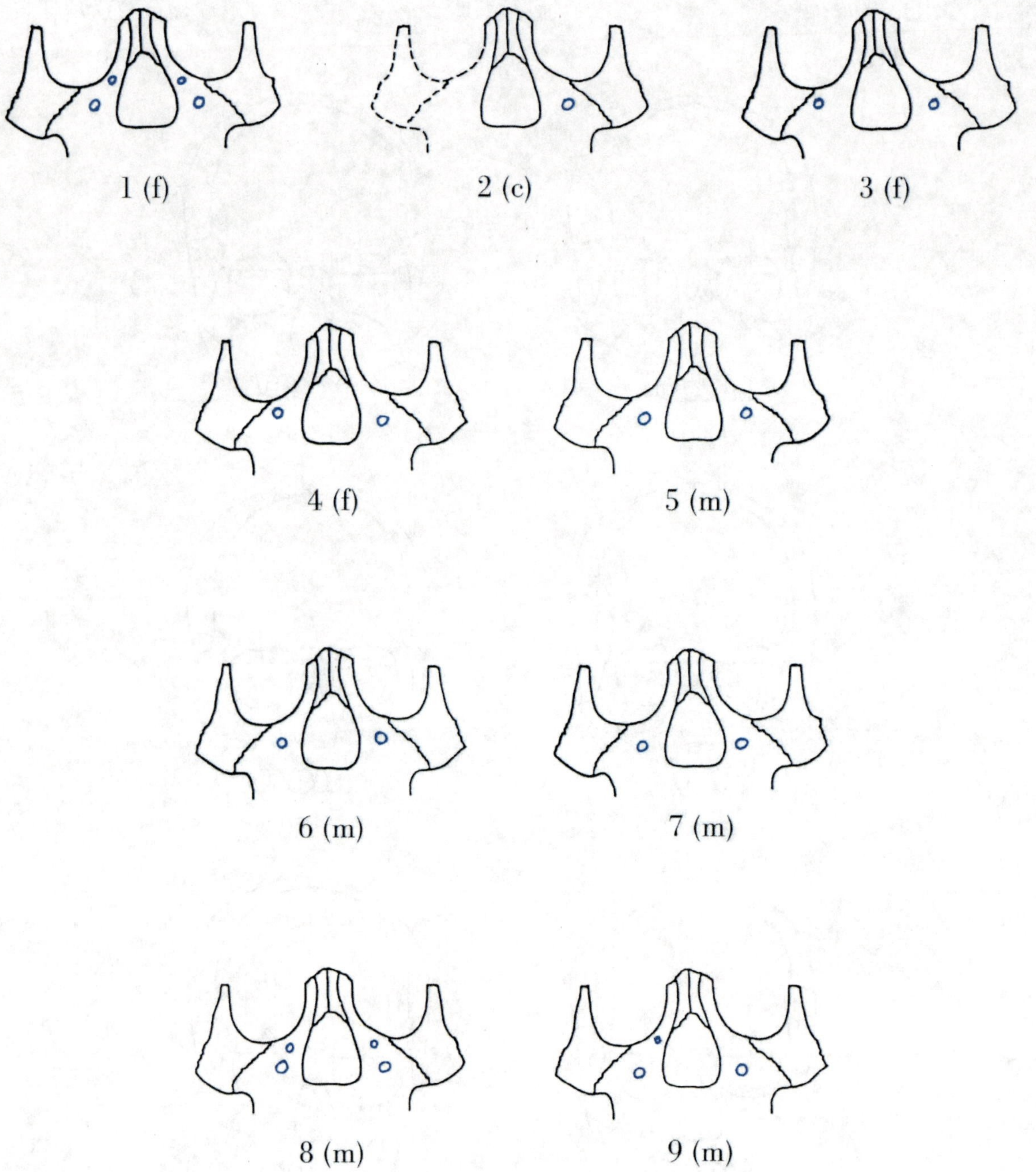

Picture 85: *Epigenetic traits – infraorbital foramina of skulls 1–9*
Legend: *dotted lines indicate places where the bone is broken off; f = female, m = male, c = child*

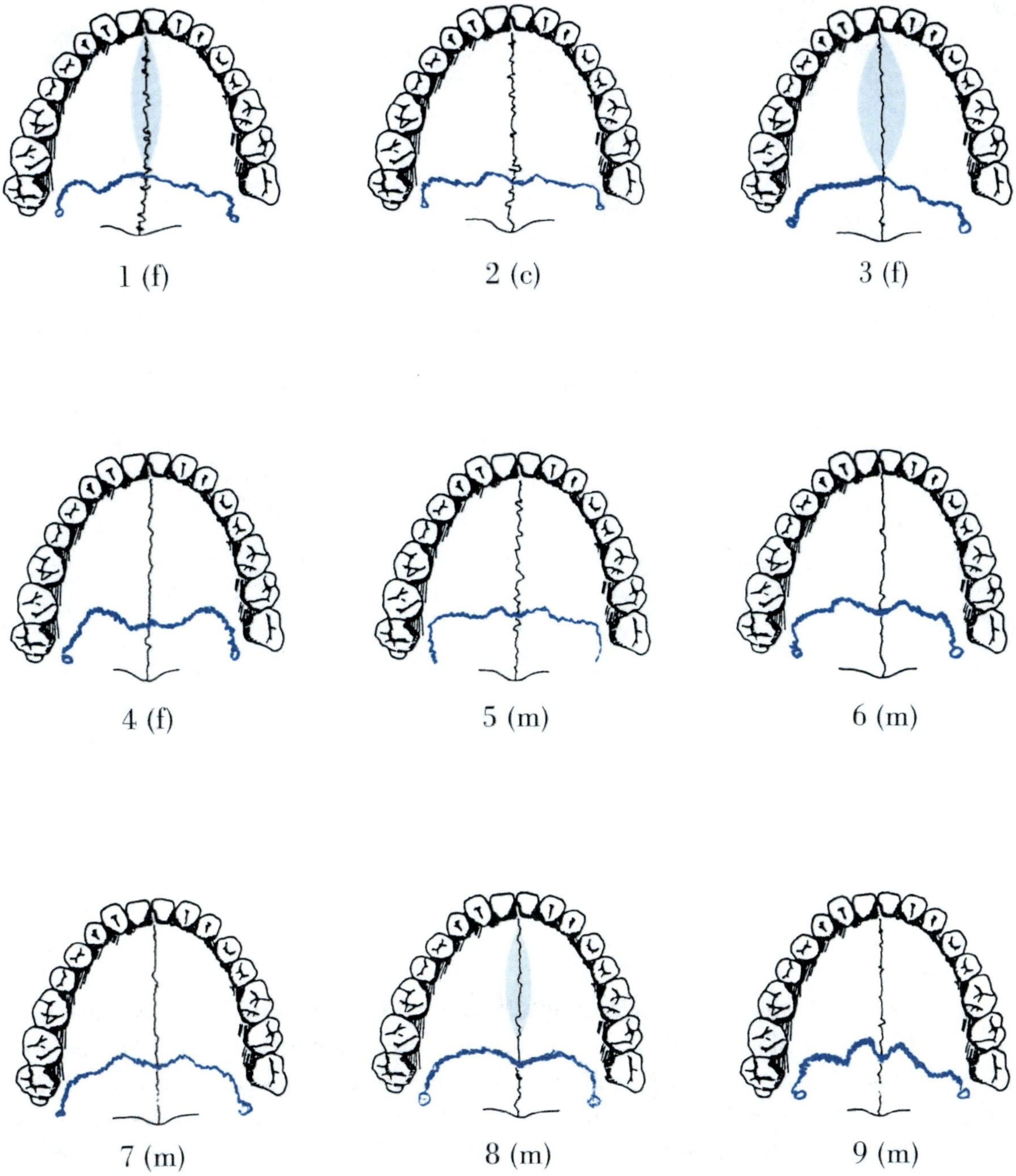

Picture 86: *Epigenetic traits – transverse palatine sutures of skulls 1–9. Alongside the medial palatine suture are the shaded palatine tori*
Legend: *f = female, m = male, c = child*

If one wishes to follow the genealogical tree, it is at this point difficult to continue with linked identification since the skull of Hermann II of Celje is missing. Because during his life the estates of the Celeians had greatly expanded, he wished to have a more distinguished place for his tomb. He established the monastery of Pleterje and was buried in the monastery church. At the transition from the 19th to the 20th century, his mortal remains were for unknown reasons transferred to the local village cemetery and buried in an unmarked grave. All attempts to gain access to his posthumous remains have proved unsuccessful. Therefore, at this juncture, the research was continued on the skulls from the end of the genealogical tree.

4.4.4.
The Family of Friedrich II of Celje (skulls 17, 11 and 18)

The last male descendant of the Counts of Celje was Ulrich II, to whom the historical data ascribe the skull with half of the face cut off, and bearing the number 18 on the back of the head. The posthumous remains of his wife Catherine Branković are assuredly not in the vault, and the same may also be said of the daughter Elisabeth, married to Hunyadi. Consequently, in searching for the kinship between the remaining skulls, an endeavour was made to reconstruct the male kinship line.

Ulrich II of Celje was killed in the year 1456, and his father Friedrich II of Celje had died only two years before this. Ulrich's skull was marked with the highest number, 18, probably because he was the last to be interred. Considering the order of burials, one may therefore anticipate that the skull of Friedrich II of Celje would be marked as number 17. In its identificational features, skull 18 belongs to an adult male aged approximately 50 years. Skull 17 bears the features of an old male skull. From the sources it is known that, among the Celeians, Friedrich II of Celje lived to the greatest age, approximately 80, to which this skull corresponds.

Friedrich II of Celje was married to Elisabeth of Frangepan, with whom he had a son, Ulrich II of Celje. By drawing conclusions from the genealogical and historical data, and by comparison with the identification results, for skulls 17 and 18 – most probably father and son – it was possible to identify the skull which corresponded to the features of the mother. In doing so, the date of her burial was also taken into consideration, later than Anna of Schaunberg and earlier than Elisabeth of Abensberg, as the expected and actual age of the skull were at about 40 years. To all these features the female skull 11 conformed.

A comparison of the facial cavities of skull 17 as father, 11 as mother, and 18 as son was made (pict. 87):

No. 17: ♂, over 60 years | No. 11: ♀, 40 years | No. 18: ♂, 50 years

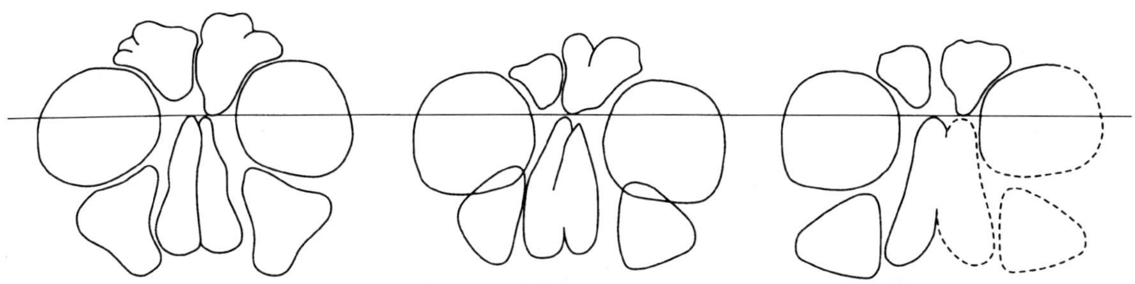

Friedrich II of Celje – father | **Elisabeth of Frangepan –** mother | **Ulrich II of Celje –** son

Picture 87: *Facial cavities of skulls 17, 11 and 18*

Commentary: the comparison indicates that skull 17 has more trapezoidal orbits, 11 rectangular, 18 also rectangular. The frontal sinuses are asymmetric in skull 17, somewhat larger and fan shaped; in skull 11, they are also asymmetric, somewhat smaller and also fan

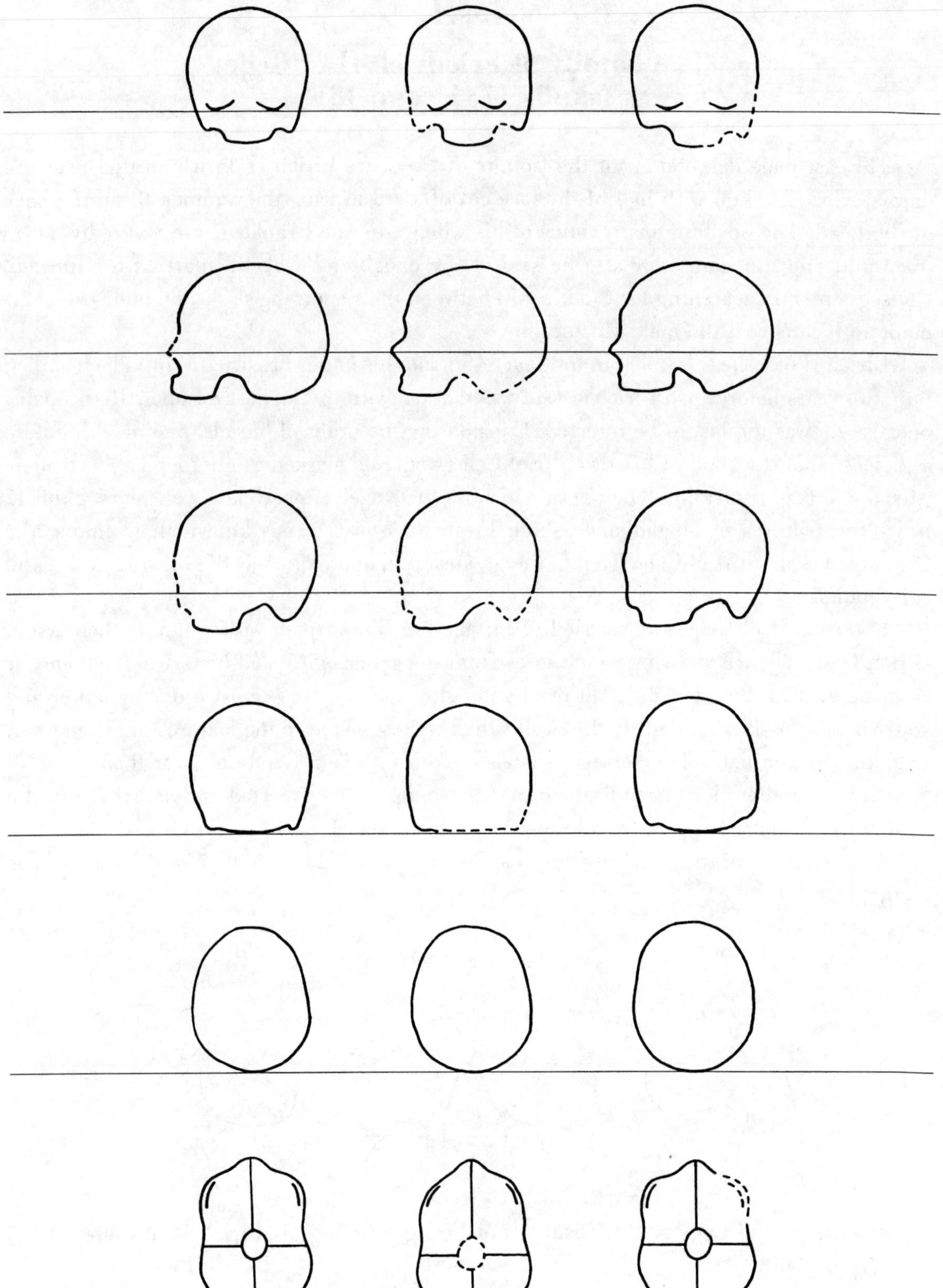

Picture 88: *Outline of skulls 17, 11 and 18 from the front, in semi-profile, from behind, from above, from below (sketched after the photographs by V. Berk)*

shaped. In skull 18 the frontal sinuses are similar to those of skulls 17 and 11, only that they are smaller. In skull 17 the maxillary sinuses are more leaf shaped, while in 11 and 18 they are triangular. In skull 17 the nasal cavities are scallop shaped, in 11 pear shaped, but for skull 18 – on account of the injury – they are more difficult to assess, although at least the right half is more pear shaped. Comparisons indicate that the supposed son inherited more of the features of the facial cavities from his mother. Skull 18 is smaller than skull 17 in size as well.

More important epigenetic traits for skulls 17, 11 and 18

- supraorbital foramina and supratrochlear notches (pict. 93)
  17 – right rounded notch and small foramen, left rounded notch
  11 – right two foramina, left rounded notch
  18 – right expressed notch, left slighter notch
- (double) infraorbital foramina (pict. 94)
  17 – left and right single
  11 – right one, left injury
  18 – left broken off, right single
- double zygomatic-facial foramina
  17 – right and left double
  11 – right and left injured
  18 – right double, left broken off
- parietal foramina (pict. 96)
  17 – right 1
  11 – right and left 0
  18 – right 1, left 1
- extrasutural mastoid foramina
  17 – left 0, right 0
  11 – left 1, right injury
  18 – left 1, right 1
- supranasal sutures (pict. 97)
  17 – right and left
  11 – right and left
  18 – complete metopic suture
- transverse palatine suture (pict. 95)
  17 and 18 cannot be compared, since skull 18 is missing entire maxilla
  11 – see picture
- marginal tubercle
  17 – not present
  11 – right visible, left injury
  18 – right deformed
- torus palatinus (pict. 95)
  17 and 11 – present
  18 – maxilla broken off

On account of the severe injury to skull 18, comparison of these epigenetic traits is not sufficiently scientifically grounded.

Among the epigenetic traits on skull 18 there are important details of the small sutural ossicle right in the lambdoid suture, and the metopic suture on the forehead. Both characteristics represent an important inherited component among the epigenetic traits. The outer contours of skulls 17, 18 and 11 were also compared (pict. 88).

Conclusion: comparison of the facial cavities of skulls 17, 11 and 18, and of certain other epigenetic traits, indicates that person 18 inherited more noted features from person 11 than from person 17.

Considering the identification, historiography, genealogy and numbering of the skull, it may be concluded that:

SKULLS 17, 11 AND 18 ARE INTERRELATED BY FAMILY. WITH A HIGH DEGREE OF PROBABILITY THEY WERE IDENTIFIED AS THE FATHER FRIEDRICH II OF CELJE (1), MOTHER ELISABETH OF FRANGEPAN (11), AND SON ULRICH II OF CELJE (18).

4.4.5.

(The Family of) Hermann III of Celje (skulls 15 and 12)

Hermann III of Celje did not have male descendants. His first wife, Elisabeth of Abensberg, reached only the age of 30–40 years. It was anticipated that the posthumous remains numbered 12 would belong to her, as it is historically argumented that her expected death date is after that of Elisabeth of Frangepan, and prior to the year 1424, when Hermann III of Celje remarried; and as it is substantiated by the epigenetic traits, which do not match any of the skulls studied, since she was not a blood relation (pict. 89).

The serious disease-induced changes on skull 12 (cribra orbitalia, pict. 57) are indicative of severe anaemia. The sieve-like perforated orbital roof also accompanies a malnourished condition to a great extent. However, since malnourishment was not frequent among the nobility, it is conjectured that the cause was anaemia; according to the sources, such characters would most frequently indicate sideropenic anaemia. It is also possible that the wife died in childbirth, which in those times – despite the best of care – was no rarity indeed.

Examination of the size of the facial cavities indicates a small, short skull with undeveloped frontal sinuses; one can observe only the small ethmoidal sinuses. Among the epigenetic traits on skull 12, only the absence of the supraorbital and double infraorbital and zygomatic-facial foramina, and one parietal and extrasutural foramen will be mentioned. More important, however, is the smaller (0.5 cm^2 large) ossicle in the centre beside the lambdae, and one of approximately the same size in the centre of the lambdoid suture.

No. 15: ♂, 30–40 years — No. 12: ♀, 30 years

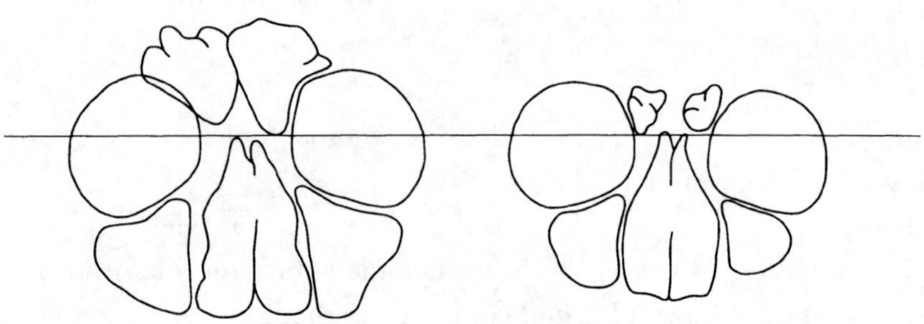

Hermann III of Celje – husband — **Elisabeth of Abensberg –** wife

Picture 89: *Facial cavities of skulls 15 and 12*

Considering the identification, historiography, genealogy and numbering of the skull

IT MAY BE CONCLUDED WITH A GREAT DEGREE OF PROBABILITY THAT SKULLS 15 AND 12 BELONGED TO HERMANN III OF CELJE (15) AND TO HIS FIRST WIFE ELISABETH OF ABENSBERG (12).

4.4.6.
(The Family of) Ulrich II (skulls 18 and 16)

Next to the posthumous remains of Ulrich II of Celje (18), one would expect to find only the still juvenile son Hermann IV of Celje, among those jointly interred from his family. The child's skull 16, following the identificational, historical and genealogical data and its number 16 (one before last to die), belongs hypothetically to the son. One may compare the epigenetic traits of skulls 18 and 16, supposedly those of the father and the son, and the latter also with the traits of skulls 17 and 11, supposedly those of the grandparents. There are no posthumous remains of the mother.

The facial cavities develop up till the end of physical growth and development. Hence, the shape and size of the sinuses in skull 16 indicate the developmental phase of a child aged 10–12 years, and therefore cannot be taken as being the completed state. Since we also do not know the epigenetic traits of the mother's skull, the outlines of the facial cavities of skulls 18 and 16 are only informative (pict. 90). They do indicate a similarity. Among the epigenetic traits on skull 16 no particularities are to be found.

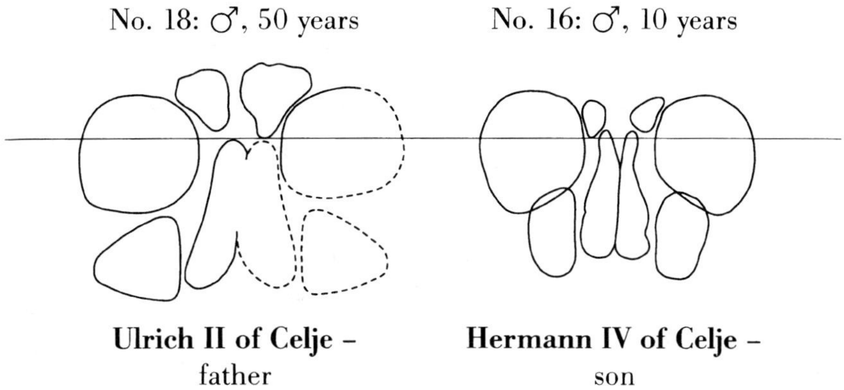

Picture 90: *Facial cavities of skulls 18 and 16*

Considering the identification, historiography, genealogy and numbering of the skull:

THE REMAINS SHOW THE PROBABILITY THAT SKULLS 18 AND 16 BELONGED TO THE FATHER ULRICH II OF CELJE (18) AND THE JUVENILE SON HERMANN IV OF CELJE (16).

4.4.7.

The Family of Count Hermann II of Celje (skulls 9, 10, 13, 15 and 17)

According to the historical sources, among those jointly interred in the family vault one would expect to find the descendants of Hermann II of Celje (interred in Pleterje), i.e. the sons Ludwig, Hermann III of Celje and Friedrich II of Celje, and possibly also the illegitimate, later legitimized, son Hermann. Epigenetic identification was performed on male skulls 9, 13, 15 and 17. Considering the identificational, historical and genealogical data and the numbering of the skulls, it was hypothetically presumed that skull 9 belonged to Ludwig (d. 1417), (then follow the female skulls 10, 11 and 12), skull 13 to the son Hermann (d. 1421), skull 15 to Hermann III of Celje (d. 1426/28), and skull 17 to Friedrich II of Celje (d. 1454). Studies were carried out to determine which of the female skulls could, following the methodological premises, possibly be that of the mother. Skull 10 proved to correspond, in accordance with the identificational, historical and genealogical data, and with the numbering.

Epigenetic identification (pict. 91):

Even a cursory glance at them will reveal that, both in shape and in size, the facial cavities of skulls 9, 15 and 17 are most similar. In all cases, the orbits are between rectangular and trapezoid, the frontal sinuses fan shaped and moderately large, the maxillary sinuses

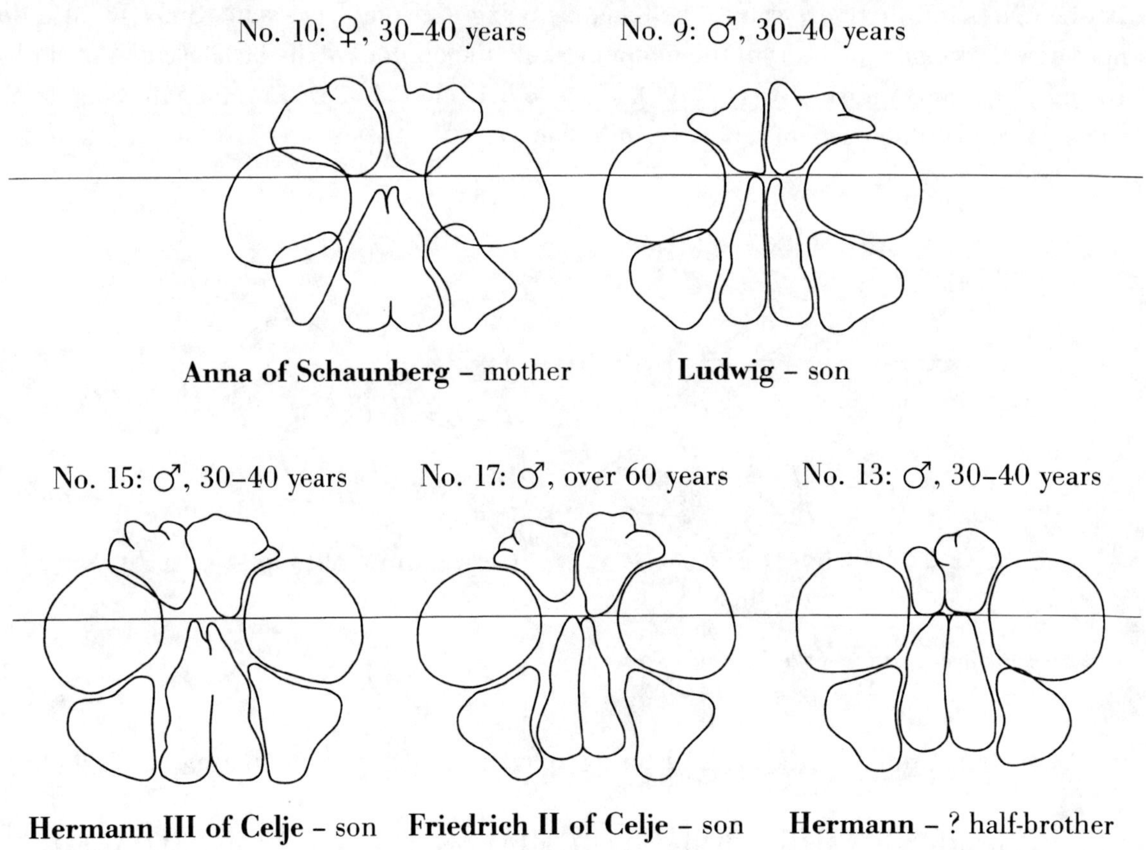

Picture 91: *Facial cavities of skulls 10, 9, 15, 17 and 13*

more triangular and of similar size, while the nasal cavities are slightly different. Skull 15 has larger nasal cavities, both being scallop shaped with serrated margins. Male skull 13 has somewhat larger facial cavities and differs most from the other three (9, 15, 17). The frontal sinuses are smaller, asymmetric and leaf shaped; the maxillary sinuses are more kidney shaped than triangular, while the nasal cavity is between scallop and pear shaped.

Female skull 10 has facial cavities which are most similar to those of skull 15. Person 15 inherited the shape of the nasal cavities after the hypothetical mother. Very similar are also skulls 17 and 9. The orbital shape is also mutually similar. The frontal sinuses of skulls 9, 15 and 17 are mutually similar, but different in skull 13. According to all conjectures, skulls 15 and 17 belong to the two brothers Hermann III of Celje and Friedrich II of Celje. Considering the advanced age of Friedrich (II) and his date of death, the oldest skull 17 is ascribed to him, while to Hermann (III), who died at the age of 35–45, the younger skull 15 is attributed.

More important epigenetic traits of skulls 17 and 15, 9, 10 and 13

- supraorbital foramina and supratrochlear notches (picts. 93, 84)
  - 17 – right rounded notch and small foramen, left rounded notch
  - 15 – right large foramen, left expressed notch
  - 9 – right and left large foramen, right slight notch
  - 10 – left foramen and slight notch, right damaged
  - 13 – right and left slight notch
- (double) infraorbital foramina (picts. 94, 85)
  - 17 – left and right single
  - 15 – left and right single
  - 9 – right double
  - 10 – left and right single
  - 13 – left and right single
- double zygomatic-facial foramina
  - 17 – right and left double
  - 15 – right and left double
  - 9 – right double
  - 10 – right and left double
  - 13 – right and left double
- parietal foramina (pict. 96)
  - 17 – right 1
  - 15 – left 1
  - 9 – left 1
  - 10 – right 1 and left 1
  - 13 – right 1
- mastoid extrasutural foramina
  - 17 – left 0, right 0
  - 15 – left 1, right 0
  - 9 – left 1
  - 10 – right 1, left 1
- supranasal sutures (pict. 97)
  - 17 – right and left
  - 18 – right and left
  - 9 – right and left
  - 10 – right and left
  - 13 – right and left
- transverse palatine suture (picts. 95, 86)
  - 17 and 15 – similarly shaped
  - 9, 10 and 13 – similarly shaped
- marginal tubercle
  - 17 – not present
  - 15 – not present
  - 9 – right and left strong
  - 10 – right and left present
  - 13 – not present
- torus palatinus (picts. 95, 86)
  - 17 – present
  - 15 – present
  - 9 – not present
  - 10 – present
  - 13 – present

Examination of the epigenetic traits investigated shows that skull 10 has one supranasal foramen, single infranasal foramen and two supranasal sutures, as in skulls 15 and 17. Skull 10 has two parietal foramina and a palatine torus. The latter is also present in skulls 13, 15 and 17.

One should not disregard the similarity between the transversal palatine sutures of skulls 10, 13, 15 and 17.

In skulls 13, 15 and 17, the shapes of the temporal bones are also very similar (picts. 65, 71). In skulls 9 and 10 they are different.

Skulls 9, 13, 15 and 17 have a convex nasal shape (slightly upturned); in female skull 10 it is straight. A shared feature of skulls 10, 13, 15 and 17 is also the palatine torus, which skull 9 does not have. The latter skull is also prognathic, while 13, 15 and 17 are orthognathic.

The shape of the skull sutures is also included among the important epigenetic traits. In skulls 13, 15 and 17 the sutural contour is unexpressed, while the sutures themselves are relatively shallow. They are similar to one another. Skull 9 has deep and more rarely winding sutures. Skull 10, however, has also well expressed sutural contours, but the meanders are compact and relatively shallow.

The above traits indicate that the inheritance is polygenic. At the same time, the features studied reveal that there are far more similarities between female skull 10 and male skulls 9, 15 and 17 than between the female skull and male skull 13. The details recorded above speak in support of the expectation that we would be dealing with the mother and three sons, while these brothers would have beside them a supposed half-brother on the father's side.

Considering the epigenetic identification, taking into account the historiographical and genealogical records, as well as the skull numberings, and also the vital fact that the skull of the hypothetical father is missing,

IT MAY BE CONCLUDED THAT SKULL 10 IN ALL PROBABILITY BELONGED TO THE MOTHER ANNA OF SCHAUNBERG, 9 TO THE SON LUDWIG, 15 TO THE SON HERMANN III OF CELJE, AND 17 TO THE SON FRIEDRICH II OF CELJE.

SINCE SKULL 13 IS SIMILAR TO THE SKULLS (OF THE HALFBROTHERS) 9, 15 AND 17, THIS SKULL PURPORTEDLY BELONGED TO HERMANN, THE ILLEGITIMATE, BUT LATER LEGITIMIZED SON OF HERMANN II OF CELJE.

4.4.8.

Friedrich II of Celje and the Illegitimate Son Hans (skulls 17 and 14)

Skull 14 is the most massive among them all. In terms of absolute values it is the largest: it has exceptionally large fan-shaped-pyramidal frontal sinuses and a strongly altered maxilla, particularly the palatine and alveolar process of the maxilla, the palatine bone and the anterior nasal aperture (picts. 61, 62). According to the historiography, the genealogical tree of the Celeians and skull numbering, it may hypothetically be ascribed to the illegitimate, later legitimized son of Friedrich II of Celje – Hans. Comparison of the facial cavities of skull 17 – the hypothetical father Friedrich II of Celje – and of skull 14 (pict. 92), in terms of shape indicates a similarity in the orbits, nasal cavities and upper maxillary sinuses. In the former case, the frontal sinuses are fan shaped, while in the latter they are pyramidal. The exceptional size of the skull, together with the facial cavities, has been explained as the result of Albright's syndrome, from which this person in all probability suffered, and which was also accompanied by acromegaly. Since, in their shape, the facial cavities display a similarity with those of the hypothetical father, it may be conjectured that the skull belonged to the illegitimate son Hans. In this case, the characteristics examined could also be similar to those of the supposed half-brother Ulrich II of Celje.

Among the epigenetic traits of skulls 14 and 17, it should be indicated that both have palatine tori, which are a family trait. Furthermore, one cannot disregard the pathological changes on skull 14, which are confirmed by macroscopic examination as Albright's syndrome or fibrous dysplasia. This benign, but long-lasting process led to the monstrous appearance of the face.

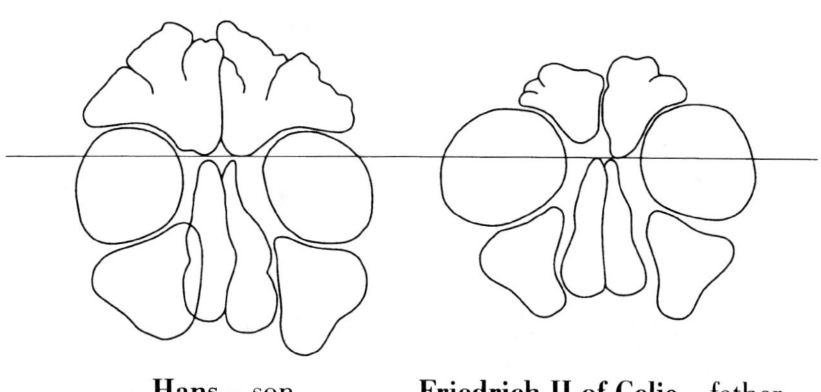

Picture 92: *Facial cavities of skulls 14 and 17*

THE SUPPOSITION REMAINS THAT SKULL 14 BELONGED TO THE ILLEGITIMATE SON OF FRIEDRICH II OF CELJE – HANS.

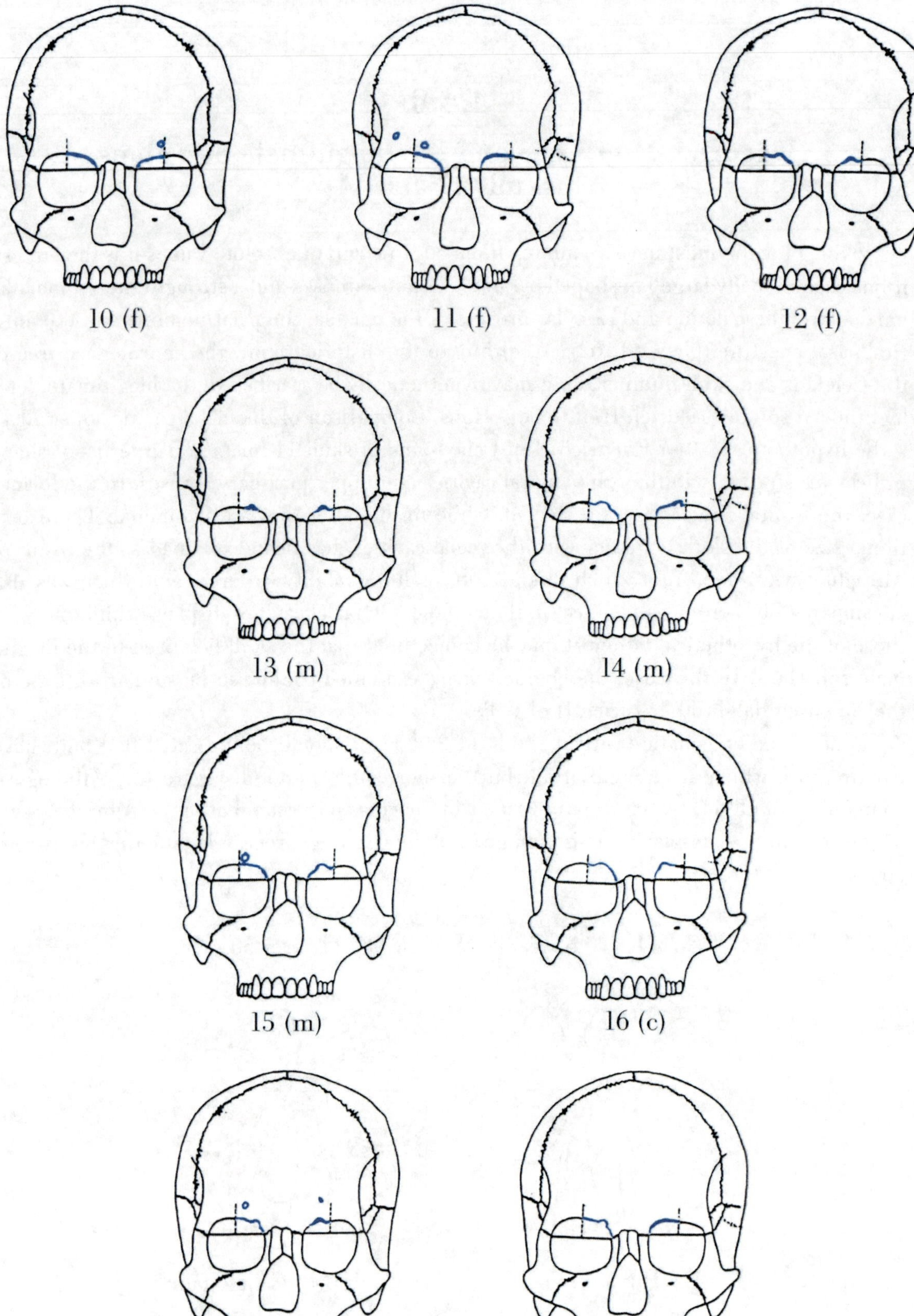

Picture 93: *Epigenetic traits – supraorbital foramina and notches of skulls 10–18*
Legend: *f – female, m – male, c – child*

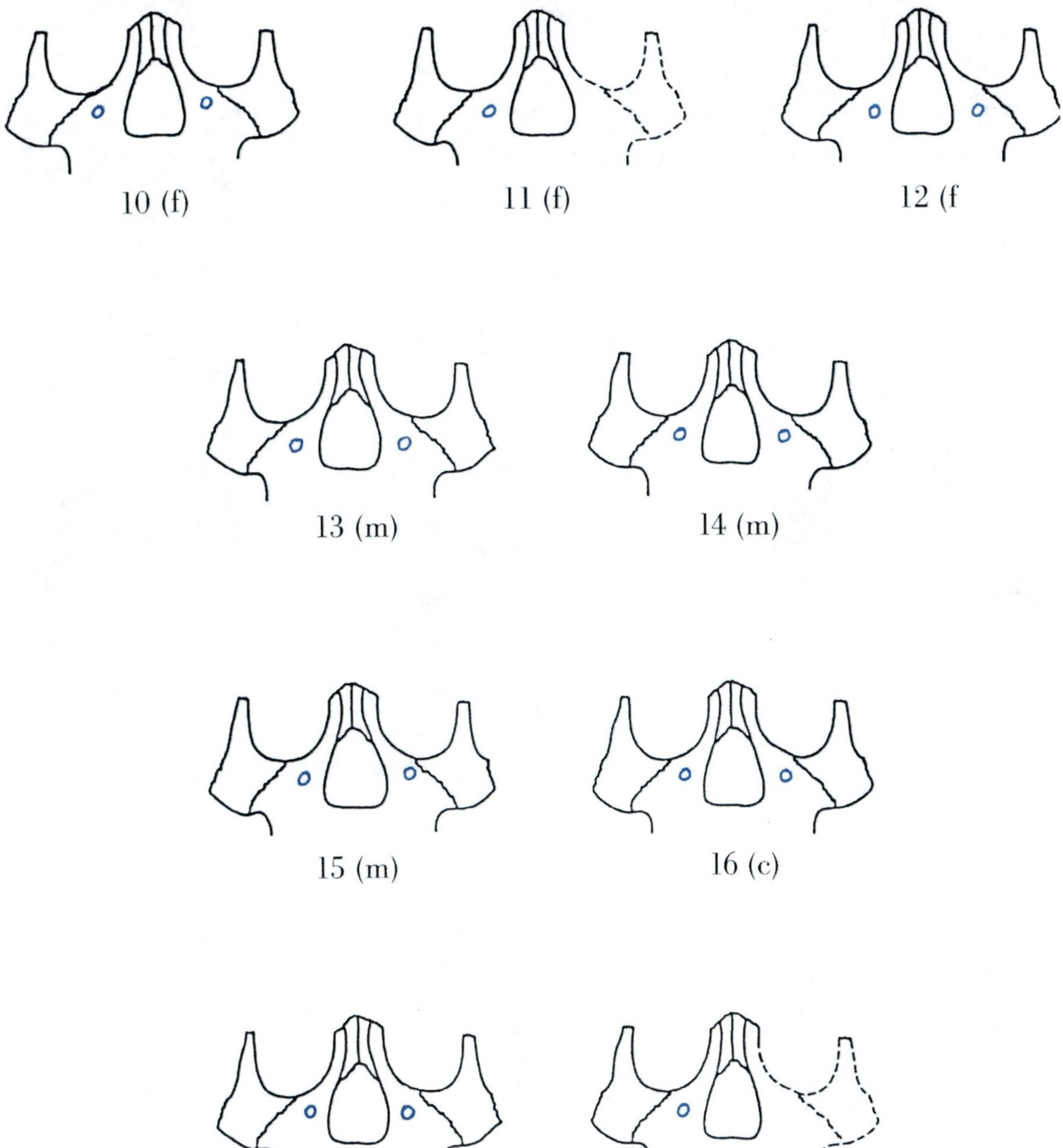

Picture 94: *Epigenetic traits – infraorbital foramina of skulls 10–18*
Legend: *dotted lines indicate places where the bone was broken off; f – female, m – male, c – child*

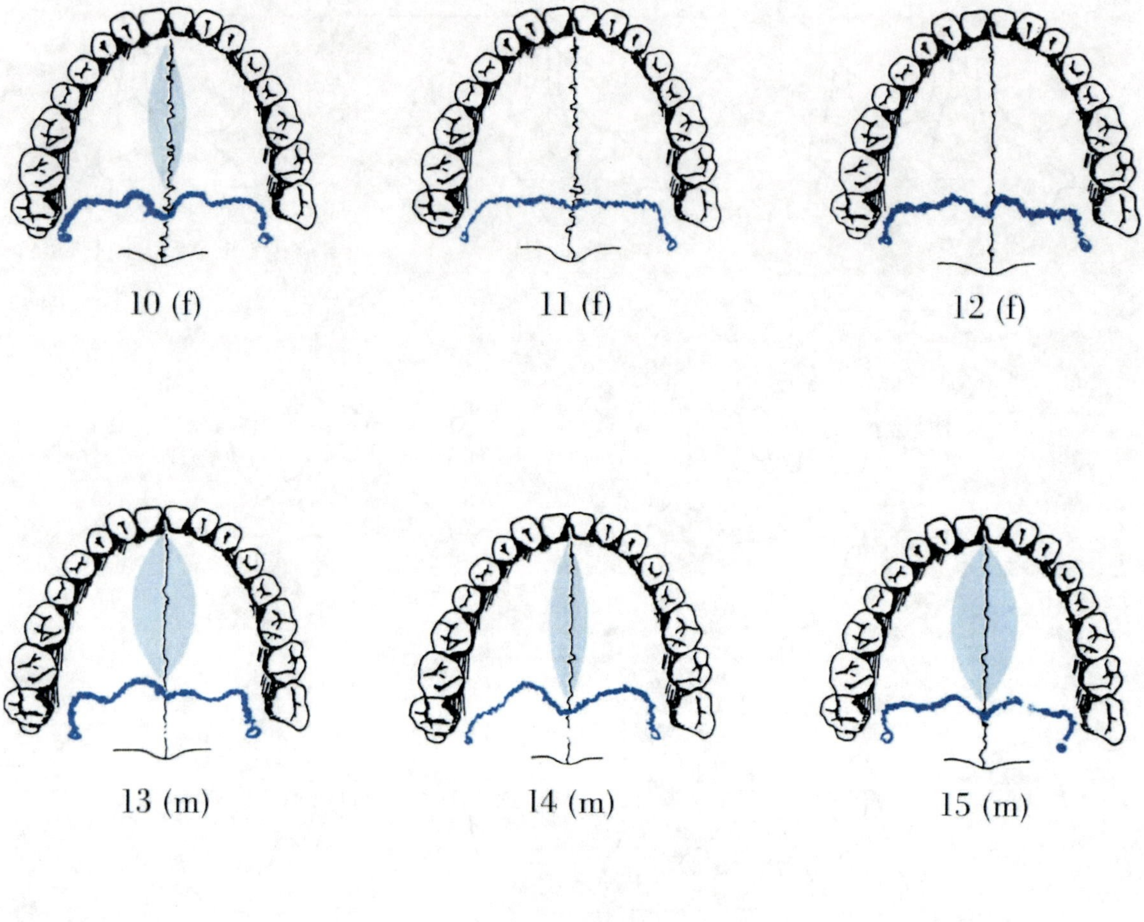

Picture 95: *Epigenetic traits – transversal palatine sutures of skulls 10–18. Skull 18 has a partially broken off maxillary palatine process on the right. Alongside the medial palatine suture are the shaded palatine tori*
Legend: *f – female, m – male, c – child*

4.4.9.
Common Epigenetic Traits of Skulls 1–18

Conclusions about the epigenetic inter-relation between parents and children cannot be drawn unless one has the posthumous remains of both parents, and yet even in this case one still needs to recall that inheritance is polygenic, that in children of the same parents different genetic combinations may occur, expressions of different genes, and that consequently the skulls of the children may also display different hyperostotic or hypostotic characteristics.

Complete comparison of the facial cavities among the18 skulls is also of importance. For the male members of the family it may be stated that the frontal sinuses are almost without exception fan shaped, the greatest deviation being observed in skulls 13 and 18. It is interesting that analyses of the frontal sinuses among the Austrians have shown that among the population that type of frontal sinus is the most widespread. In the majority of the male skulls studied, the maxillary sinuses are of triangular shape. The orbital openings in our case most frequently resemble a rectangular-trapezoid shape, none of them being rounded or ellipsoid. Also the nasal openings are as a rule almost all scallop shaped. Such a high degree of uniformity in the shape of the facial cavities amongst the male members, i.e. blood relations, is certainly not coincidental. In addition to the shape, the size of the facial cavities is also important: throughout the century of family burial under study, the frontal sinuses displayed a moderate trend towards reduction in size. Although this trend does not hold for the orbital surfaces, it is nevertheless more expressed around the nasal cavities.

The facial cavities of 6 female skulls – i.e. non-blood relations – have frontal sinuses which are more leaf-fan-shaped, with the exception of skull 12, in which they had not developed. The maxillary sinuses are closest to triangular in shape. The shapes of the orbital openings are between rounded and rectangular, while the nasal cavities in three cases are pear shape, in the other three scallop shape. All the female facial cavities are, as expected, smaller than the male ones. In general, it may be concluded that the sons buried in the family vault inherited the shapes of the facial cavities to a great extent after the fathers. As for the daughters who were married elsewhere, there is no information on this feature. The size of the facial cavities also provides an indirect indication of the height of the upper part of the face or head, and hence also the person's height. Among the women, the one with the shortest face was person 12 (Elisabeth of Abensberg), followed by person 1 (Didmund of Walsee), then person 3 (Adelaide of Ortenburg), person 11 (Elisabeth of Frankopan), person 4 (Catherine Kotromanić, while the longest was that of person 10 (Anna of Schaunberg). Overall, with regard to the facial cavities of the skulls examined, one may suppose that, over the century, the descendants diminished in height. They married women of small stature, and this in the longer term affected physical growth in the relatives. The one exception is skull 14, which was affected by the disease described earlier.

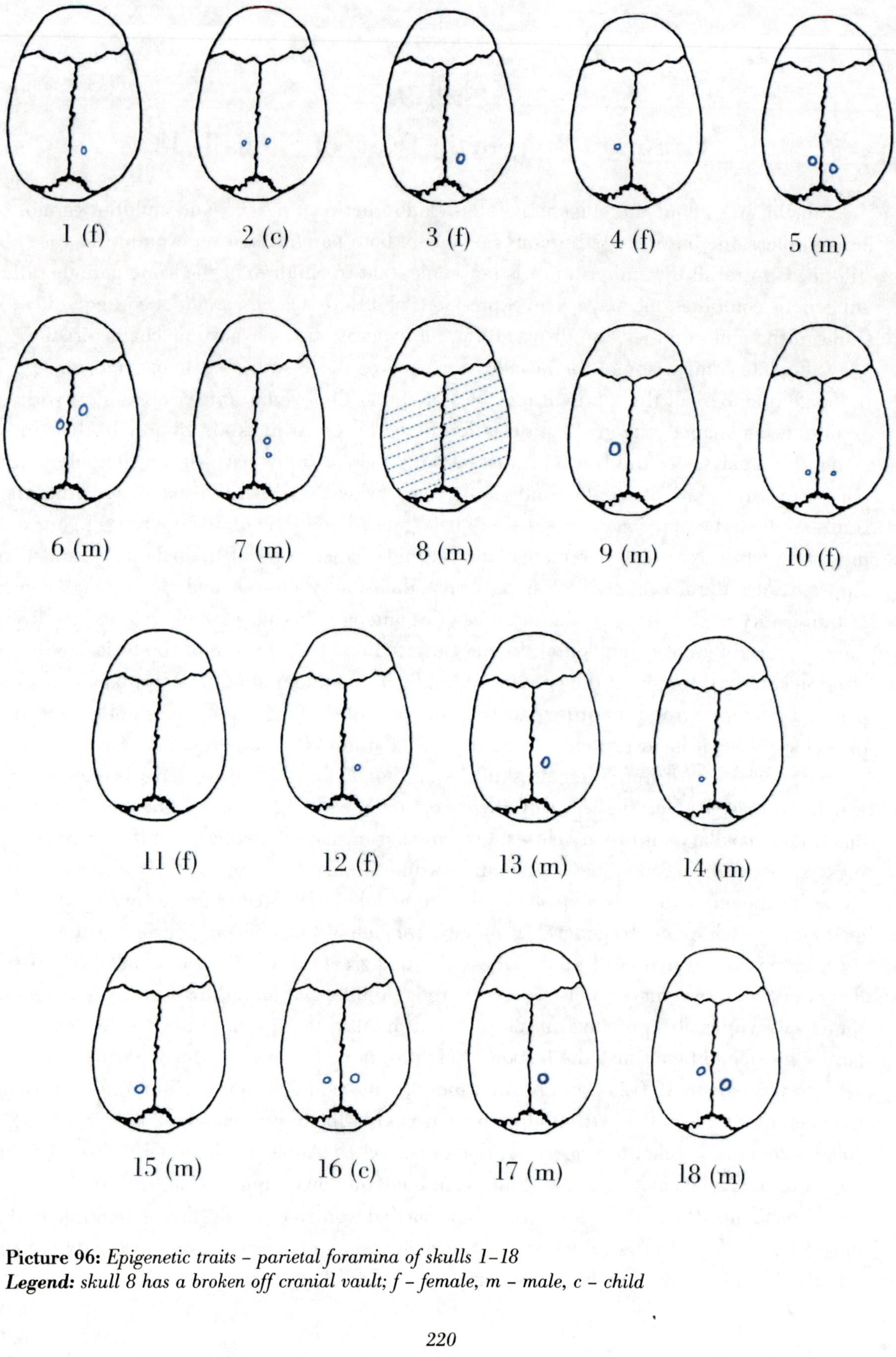

Picture 96: *Epigenetic traits – parietal foramina of skulls 1–18*
Legend: *skull 8 has a broken off cranial vault; f – female, m – male, c – child*

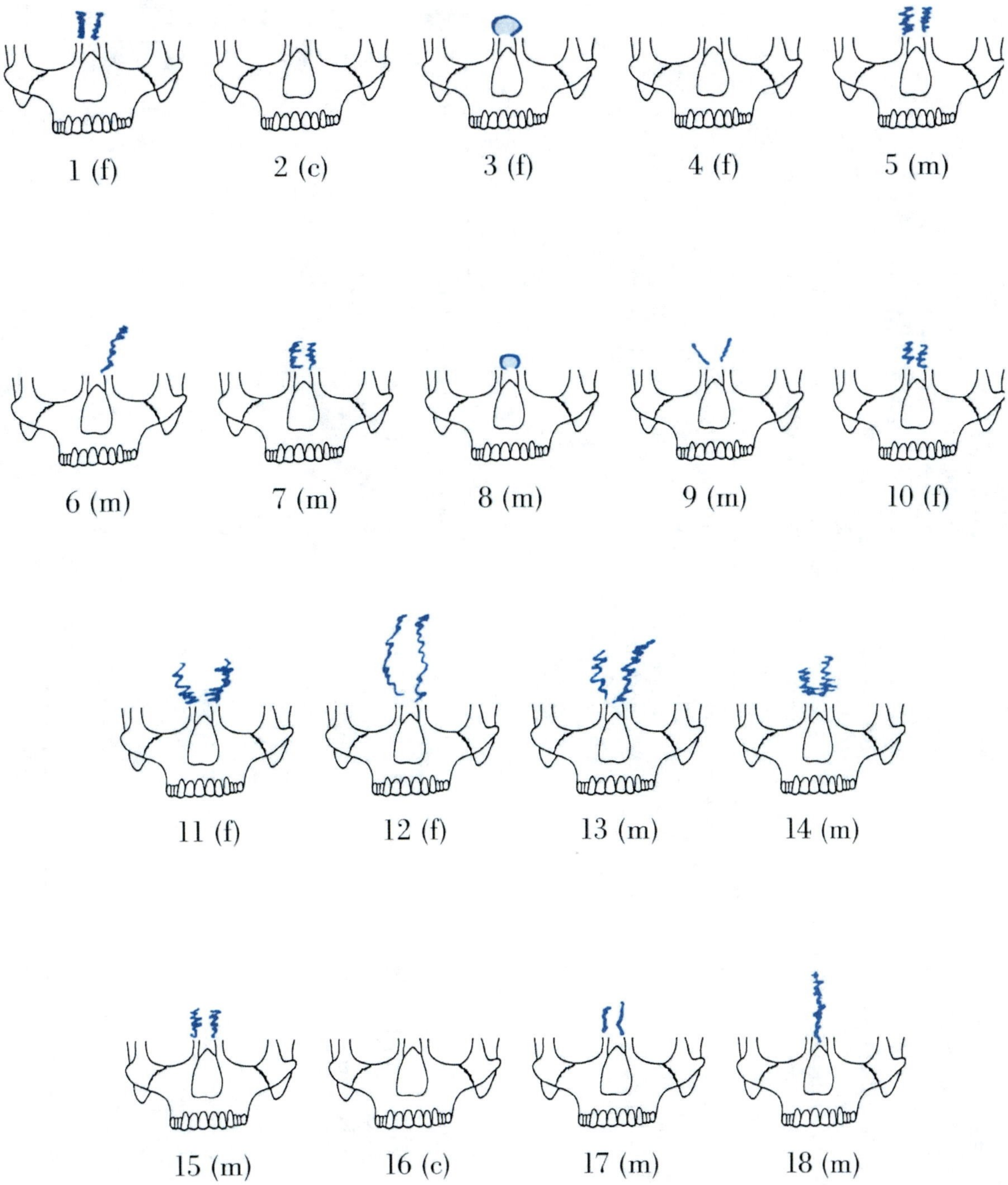

Picture 97: *Epigenetic traits – supranasal sutures of skulls 1–18*
Legend: *skulls 3 and 8 have rare ossicles above the nasal bone; f – female, m – male, c – child*

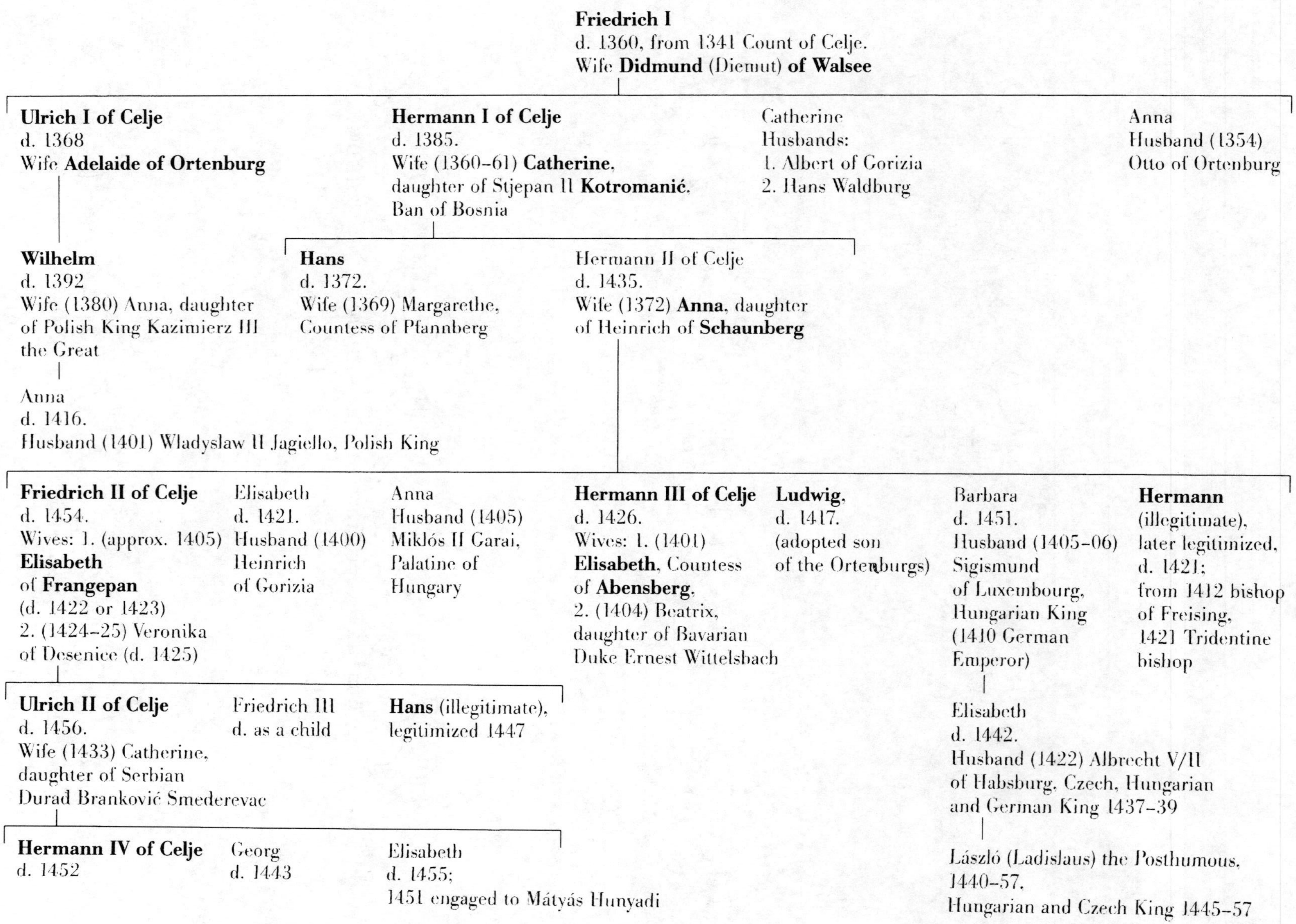

Picture 98: *Genealogical presentation of identified members of the family of the Counts of Celje*

4.5.
Conclusions from the Historiographical and Identificational Results

THE RESEARCH, WITH A HIGH DEGREE OF PROBABILITY, LED TO THE FOLLOWING NAMES (pict. 98):

Skull **1** – **Didmund of Walsee** (died 1353 or 1354)

Skull **2** – **Hans** (died 1372)

Skull **3** – **Adelaide of Ortenburg** (died 1391)

Skull **4** – **Catherine Kotromanić** (died before 1408)

Skull **5** – **Friedrich I** (died 1359 or 1360)

Skull **6** – **Hermann I of Celje** (died 1385)

Skull **7** – **Ulrich I of Celje** (died 1368)

Skull **8** – **Wilhelm** (died 1392)

Skull **9** – **Ludwig** (died 1417), (probably, but no skull of the father)

Skull **10** – **Anna of Schaunberg** (died before 1396)

Skull **11** – **Elisabeth of Frangepan** (died 1422 or 1423)

Skull **12** – **Elisabeth of Abensberg** (died before 1424)

Skull **13** – **? Hermann**, legitimized son of Hermann II of Celje (died 1421), supposedly

Skull **14** – **Hans**, legitimized son of Friedrich II of Celje (died after 1447), supposedly

Skull **15** – **Hermann III of Celje** (died 1428) (probably, but no skull of the father)

Skull **16** – **Hermann IV of Celje** (died 1444) (probably, but no skull of the mother)

Skull **17** – **Friedrich II of Celje** (died 1454) (probably, but no skull of the father)

Skull **18** – **Ulrich II of Celje** (died 1456)

Picture 99: *Friedrich I, portrayed according to anthroposcopic, anthropometric and facial shape characteristics, determined with a high degree of probability from his skull (5). Work of the Academy painter Rudi Španzel (1999)*

4.6.
Genealogical Results

Research studies frequently lead to unforeseen discoveries and results. In this study of the Counts of Celje, all the extant genealogical records were consulted. Subsequent work also led to assessment of the genealogical continuation through Barbara of Celje, daughter of Hermann II of Celje. It was surprising to discover that the kinship traits, already 20 generations after her, had remained uninterrupted. In the Middle Ages, it was considered that, with the death of the male descendants, the lineage dies out.

The discovery of descendants from the line of Barbara of Celje (of Cilli) provided this study with new research points. The genetic reference material stimulated a genetic study: the possibly isolated DNA from the skulls studied could be compared with that taken from the still living descendants. This would provide a final answer as to whether the skulls studied did actually belong to the Counts of Celje. Furthermore, the discovery that the lineage after Barbara of Celje was uninterrupted was invaluable, since this would prove decisive in the case of successful DNA isolation. In fact, mitochondrial DNA is inherited through the female line and is far more resistant than the nuclear DNA; it does not recombine, and is used for paleogenetic studies. Of additional interest is that, from the inherited material of the descendants through Maria Luiza, Princess of Hessen and Kassel (b. 9. 5. 1814, d. Hohenburg 28. 7. 1895), of the 16th generation after Barbara of Celje, inherited material was taken in 1994 for identification of the Russian imperial family of the Romanovs, murdered during the October Revolution in 1917. These facts confirm the conclusion that, so to speak, the entire nobility of Europe was kindred by a closer or more distant blood relation.

Picture 100: *Friedrich II of Celje, portrayed according to anthroposcopic, anthropometric and facial shape characteristics, determined with a high degree of probability from his skull (17). Work of the Academy painter Rudi Španzel (1999)*

Representatives of the unbroken female line through Barbara of Celje:

1. Barbara of Celje (b. in Celje around 1392, died in Melnikov on 11. 7. 1451)
2. Elisabeth of Luxembourg (b. in Prague, 28. 2. 1409, d. in Ofen 19. 12. 1442)
3. Elisabeth of Austria (b. around 1436, d. in Poland 30. 8. 1505)
4. Barbara of Poland (b. 15. 4. 1478, d. 15. 2. 1534)
5. Christine of Saxony (b. 25. 12. 1505, d. 15. 4. 1549)
6. Christine of Hessen (b. 29. 6. 1543, d. 13. 5. 1604)
7. Anna of Schleswig-Holstein-Gottorp (birth and death details unknown)
8. Anna of Ostfriesland (b. 23. 6. 1601, d. 15. 2. 1634)
9. Anna Maria Dorothea, Duchess of Mecklenburg (b. 1. 7. 1627, d. in Halle 11. 12. 1669)
10. Magdalena Sibylla, Duchess of Sachsen-Weissenfels (b. 2. 1648, d. in Gotha 7. 1. 1681)
11. Anna Sophia, Duchess of Sachsen-Gotha-Altenburg (b. 22. 12. 1670, d. in Rudolfstadt 28. 12. 1728)
12. Anna Sophia, Princess of Schwarzburg-Rudolfstadt (b. 9. 9. 1700, d. in Romhild 11. 12. 1780)
13. Charlotte Sophie, Duchess of Sachsen-Saalfeld-Cob. (b. 24. 9. 1731, d. in Schwerin 2. 8. 1810)
14. Sophia, Duchess of Mecklenburg-Schwerin (b. 24. 8. 1758, d. in Sorgenfrei 29. 11. 1794)
15. Louise Charlotte, Danish Princess (b. 30. 10. 1789, d. in Copenhagen 28. 3. 1864)
16. Marie Louise, Princess of Hessen-Kassel (b. 9. 5. 1814, d. in Hohenburg 28. 7. 1895)
17. Bathildis, Princess of Andalt-Dessau (b. 29. 12. 1837, d. in Nachod 10. 02. 1902)
18. Bathildis Marie Leopoldine Anna Auguste, Princess of Schaumburg-Lippe (b. 21. 5. 1873, d. in Arolsen 21. 4. 1962)
19. Helene, Princess of Waldeck and Pyrmont (b. 12. 12. 1899, d. in Rasted 18. 2. 1948)
20. Eilika Stephanie Elisabeth Tekla Juliana, Duchess of Oldenburg (b. 2. 2. 1928)

20th Generation

- Eilika Stephanie Elisabeth Tekla Juliana, Duchess of Oldenburg, was married on 10. 8. 1950 to Emih VII of Leningen; they have 4 children, named:
- Melita of Leningen (b. 1951)
- Karl of Leningen (b. 1952)
- Andrej of Leningen (b. 1955)
- Stephanie of Leningen (b. 1958)

The latter should be the bearers of the same genetic record as that of Barbara of Celje.

The Oldenburg – Leningen family is acquainted with the research. They were also invited to the exhibition on the Counts of Celje at the Slovenian National Museum. In the case of positive isolates of mitochondrial DNA from the skulls studied, their reference DNA samples could be compared (5 hairs are already sufficient).

Picture 101: *Ulrich II of Celje, portrayed according to anthroposcopic, anthropometric and facial shape characteristics, determined with a high degree of probability from his skull (18). Work of the Academy painter Rudi Španzel (1999)*

4.7.
Art History Results

The Slovene people have always wished to know the images of prominent people who have lived in the Slovene lands. The distant times of the life and activity of the Counts of Celje have deprived us of their original portrayals. This is why, as part of this study, I have also set myself the aim of studying all the shaping characteristics of their faces and skulls, in order to give the artist a sound basis for the representations. With a high degree of probability, it was possible to identify the skulls of Friedrich I (no. 5), Friedrich II of Celje (no. 17) and Ulrich II of Celje (no. 18).

The painter was provided with the following details concerning:

- approximate age,
- shape and size of their skulls,
- height of the upper facial part,
- height of forehead,
- breadth of face,
- shape of orbits, which determines the shape of the lower and upper eyelids,
- expression of supraorbital arches,
- inclination of forehead,
- prominence of glabella,
- breadth of nasal root,
- shape of nose (but not length, which cannot be determined posthumously),
- breadth of cheekbones,
- maxillary cheekbone prominence (processus zygomaticus maxillae),
- Campers' angle, which determines whether the upper teeth are normally positioned
- prominence of muscle attachments, which indicate a more or a less muscled face,
- degree of sexualization of these skulls, which indicates whether the person's facial features were highly masculine, moderately or slightly developed,
- and several other details.

It is impossible to determine the numerous details which influence the facial appearance. One cannot say whether the person had considerable subcutaneous tissue, and hence a fuller face, or thinner. Likewise, one cannot predict the quality of an individual's skin, i.e. the tonus – whether the skin was smooth or wrinkled, possibly pockmarked, etc.; similarly with specific marks on the face, such as warts, scars, and pigmental indicators. Also, the colour of the eyes, hair and skin is the fruit of the artist's imagination, as are the hair, beard and similar features. A great problem in attempting to obtain the most authentic possible portrayal was posed by the absence of the lower jaw. The artist has resorted to using the growth of the beard, which covers the actual length of the lower jaw. The painter was accorded all artistic freedom, while respecting all the known parameters.

Thus the portraits were created of Friedrich I (pict. 99), Friedrich II of Celje (pict. 100) and Ulrich II of Celje (pict. 101).

5.
DISCUSSION

The purpose of the research was to identify the skulls and to study their family (blood) relationship; in this, numerous dilemmas were encountered.

5.1.
History and Genealogy

The existing genealogical trees indicate the findings hitherto concerning the Counts of Celje. In them, the biographical details are fragmentarily collected. It is certain that the historians, specialists on the Middle Ages, were mainly concerned with political events and with the economy. Biographical information was less important. If this does not hold to such an extent for the male members of the family, it is certainly true for the females acquired through marriage. Apart from certain possible years of betrothal or marriage, certain years of death, and the amount of dowry, it is not possible to glean anything from the records.

While consulting the register extracts, I was able to discover certain new biographical data; e.g. the expected lifespan of Countess Catherine Kotromanić, who had hitherto been known only from the marriage date. Especially valuable was the registry entry of the letter of betrothal between Hermann III of Celje and Beatrix of Bavaria dating from the year 1424, for whom the Slovene source (Grafenauer 1988) had given the year of marriage as 1404. With this new information, it was possible with a high degree of probability to exclude Beatrix from those jointly interred in the family vault, since her husband had died only a few years after this record, and rich widows customarily swiftly remarried.

In dealing with the biographical dilemmas concerning Hermann IV of Celje, in addition to the contradictory data of Krones and Grafenauer, new research was found on the genealogy of the Celeians in the study of 1991, in which Dr. Thomas published his birth and death details, and also those of his sister Elisabeth. This information made it possible to carry out with greater certitude the comparison between the identificatory and historical results, to determine the age of Hermann IV of Celje, and to exclude Elisabeth's posthumous remains from the family vault. Otherwise, from the historiographic records, it was not possible to glean any new facts.

In relation to the genealogy, the fact that also the female descendents are transmitters of the inherited genetic material – and that the lineage after Barbara of Celje had remained unbroken – could have been already known from before. For the research, this information was a surprise, and a stimulus for genetic analysis.

5.1.1. Numbering the skulls

Without an understanding of the principles behind the numbering of the skulls, it would be harder to elaborate the results. The working hypothesis, that the numbering does not follow the chronology of date of decease, has been confirmed.

The numbering of the skulls customarily does follow the time order of death dates, although with separate interments for male and female, or child, members of the family. We, however, are not acquainted with any medieval written sources on feudal principles regarding interment in family tombs. In the present study, the blood relatives were clearly buried separately, i.e. the males were buried separately from their hereditary wives. With regard to child's skull 2 (Hans), interred among the women, it would be expected that the children were normally buried with the mothers; however, from child's skull 16 (Hermann IV of Celje), it is evident that he was buried among the male members of the family relatives. Is this merely a chance, or did the family at that time decide that the young male offspring, prematurely deceased, nevertheless belonged among the dynasty in the vault? At that time, Hermann's father, Ulrich II of Celje, had also lost his son Georg, and was left with only the somewhat older daughter Elisabeth. Perhaps he may have had the presentiment that he would have no further male heirs, and therefore had his son Hermann IV of Celje buried among the adult males.

The research also confirmed the above with the names of those jointly interred. These dilemmas are reflected upon in the subsection titled, Identification, in the section Discussion.

What remains unclarified is the number of skulls – 19. Did this occur by mistake? One would otherwise expect to find among the buried two more small children, i.e. Friedrich III, son of Friedrich II of Celje, and Georg, son of Ulrich II of Celje. The skulls were probably not preserved on account of the fragility of the thin cranial ossicles. Here the further question arises of whether, perhaps, any of the girls also died in childhood. And if so – which would hardly be surprising, given that during the period in which the family vault existed there were eight women married to Celeian counts between 1354 and 1456 – then where are the posthumous remains? Were the possibly small children buried outside the vault? Including Friedrich III and Georg. Maybe it is for this reason that these two child skulls were not preserved. Was the age of 10 the time boundary after which the children were permitted to be interred with the adults?

Anton Zupančič, the schoolmaster of Latin, was clearly following some instructions when numbering the skulls. On the coffins from out of which he had taken the skulls, it was most probably written who was buried there, or else the burial date would have been given; in short, some detail which would have directed him. Otherwise the identification would not have matched the genealogical record and the number of skulls almost perfectly.

Despite Zupančič's praiseworthy care, nevertheless the question recurs as to why they did not also remove the mandible from the coffin. The sources do, indeed, relate that the vault was opened after the second fire in St Mary's church, which would indicate urgency and haste. It may be conjectured, however, that the fire could not have reached down to the vault beneath the stone paving, since there are no traces of fire on the skulls. Skulls without the mandible give an incomplete impression of the appearance of the face. The close example of the Habsburgs conveys to us a great deal about inherited facial features and the importance of the mandible in this respect. Since we do not have the mandibles of the Counts of Celje, it will never be possible to reliably reconstruct their appearance. Anatomical reconstructions can differ from the real appearance of the individual, if, of course, it differs from the anticipated.

5.2.
Identification

The identification was carried out independently of the historical and genealogical expectations. It was only after the skulls had been identified by gender, age, and by use of the epigenetic method, that a comparison was made between the historical and genealogical data, and the sutures on the skulls. The possibility of comparing the skulls, which was provided by various scientific methods, was conjointly used as a control mechanism for verifying the results. On numerous occasions, the diverse methods functioned in complementary fashion, supplementing one another, elucidating or eliminating. They led to broadranging reflection, objective conclusion, and critical judgement. Not infrequently, the work required mastering of new knowledge, supplementary literature, consultation with experts from other disciplines, etc.

5.2.1. Gender

Determining gender only by skulls according to the reports of different authors is not wholly reliable. Although the extensive forensic medical and anthropological methodology had in general led to indisputable results, in the examination of all the skulls the anthropometric indices were also calculated, since they determine gender dimorphism. In addition to the standard methods, non-metric methods were also used in determining the thickness of the calvaria visible from the radiographs – which has its life-regulated principles of change by gender, the shape of the dental arch and the occipital apertures, etc. Doubts about gender could thus be clarified. Particularly important was the index of sexualization.

In the children's skulls, aged approximately 10 years, sexual dimorphism is still not clearly developed. Therefore, in the case of these two skulls, gender and identity were determined within the context of the entire research: establishing kinship, taking into consideration the historical and genealogical records and the skull numberings, i.e. the dates of death. It may be recorded with a great degree of certainty that both skulls are of the male sex. This could be confirmed or refuted only by genetic analysis of the nuclear DNA. In the study, the earlier mentioned results of paleogenetic analysis were unable to isolate either the pure DNA (non-contaminated) or the mitochondrial DNA sequence, which is preserved for a long time and of which there are several in the cell. The nuclear DNA, which alone contains the gender chromosomes and can provide the answer as to gender, is not preserved altogether. Considering the age of the bone material and the sensitivity of the biological substance, in the case of these skulls the results with modern knowledge are final.

5.2.2. Age

The earlier literature does report exhaustively on the determination of age by skull with regard to obliteration of the endo- and ectocranial sutures, and puts forward the possibility of relatively precise determination of age thereby; the recent literature, however, points to the lesser reliability of this method. Forensic dentistry may also be a valuable aid, as long as the dentition is preserved. Indirect conjecturing via intact or atrophied dental alveoli may be only orientational, since it is known from practice that an elderly person may still retain relatively healthy teeth, while those of a youth may be worn out. In the research, on account of the fragmentation of the material and the loss of the dentition, there was no other choice available than to determine the age mainly by anthroposcopy and radiography following the endo- and ectocranial obliteration of the cranial sutures.

In studying the skulls, uniform criteria were applied for determining the age, based on the most recent literature, and in the adults the age was determined approximately by decades. It should not be forgotten that the skull changes little from the time of reaching sexual maturity up to the cessation of hormonal activity, i.e. from the age of 20 until approximately 50. The closing of the cranial sutures is individually conditioned, and therefore the theoretical proposals are only fair estimates. In determining the age, reference was also made to the degenerative and pathological changes: wearing down of the maxillary joints, porosity of the cranial bones, and so forth. The thickness of the calvaria was also taken into account: in youth, this is of equal thickness in both sexes, while with ageing it becomes thicker at the front in females, and at the back in males. The more intensive impressions of the cranial veins on the inner side of the calvaria, the deep arterial grooves of the medial meningeal artery, and the more numerous Pacchionian bodies served as indirect indicators for the older persons. At the same time, it can be recorded that in the study of these skulls, in many cases, the highly pronounced, otherwise physiological Pacchionian bodies were noticed, and also the more strongly prominent diploid veins, which may be regarded as a family trait and an indirect mark of kinship. On account of this feature, for differential diagnosis on skull 6 even computer tomography was carried out in order to reliably exclude a disease process.

Earlier damage to the skull, and the fragmentation of the material in general, were additional obstacles in determining the age.

The forensic-dental method for determining age could be used fully only for the two child skulls, 2 and 16. By examining the milk teeth and second teeth dental cavities the age could be assessed. Orthopan tomography proved not to be useful, since such images are standardized only for the living; consequently, only classical X-rays could be used.

In the two child skulls it was possible to assess the age also by the development of the frontal sinuses, as is described in the methodological part of the study, under the epigenetic method.

5.2.3. Diseases

Determining the pathology only by skulls is not complete, since many diseases do not leave traces on them. Hence, the results obtained reveal only the pathology in the posthumous remains examined. In general – with the exception of two cases – the causes of death were not found. Skull 18, with the smoothly excised half of the face, indubitably testifies to the cause of death. Skull 12 has both orbital roofs with sieve-like perforations, i.e. cribra orbitalia, which might represent the cause of death in the thirty-year-old woman. Whether the cause lies in severe anaemia or in possible undernutrition, one may only conjecture. It would, however, be expected that the nobility would not have suffered from undernourishment. It would also be possible that she might have died in childbirth, when an anaemic condition may be substantially aggravated. In this case, however – the skull being attributed to Elisabeth of Abensberg – there is no record of her having given birth. It is, therefore, possible that her condition of illness may actually have been responsible for her death and her inability to bear children. Otherwise, on the other skulls, no signs of tumours or of cancerous metastases were found, apart from two benign osteomas.

One of the toughest challenges was identification of the disease related changes in skull 14. The male person was suffering from a particular type of slowly advancing benign growth, i.e. fibrous dysplasia, and furthermore one of the three types of this – Albright's syndrome. With the establishment of this diagnosis, numerous dilemmas were clarified. Since the fibrous dysplasia is also associated with hormonal disturbances – and in person 14 it clearly involved acromegaly – one may explain why his skull was so incomparably larger than the others. The similarity in shape of the facial cavities with those of the father is, of course, not coincidental; the difference in size, however, is the consequence of the above-mentioned disease. In this disease, the ageing processes are accelerated, and it is therefore possible that the age by skull had been overestimated. Had Hans been interred before his half-brother Hermann III of Celje, who died in 1428, and is this why his skull is recorded as no. 14? If this is the case, then one may expect that he was legitimized by his father, Friedrich II of Celje, only posthumously in 1447, as is recorded in the sources. Could it be possible that the year 1447 is not accurate, and that Hans actually died in 1427? That the chronicler made an incorrect entry, yet the order of numbering – 14 and 15 – nevertheless holds?

On the basis of the research, it may be concluded that skull 14 – greatly increased in size by the acromegaly that accompanies Albright's syndrome – is that of the illegitimate, later legitimized son of Friedrich II of Celje, who may have died just before Hermann III of Celje, and therefore bears the number 14; and that, in some way, the order of numbering had again been reversed, as with skulls 6 and 7.

Particularly close attention was paid to examination of the younger skulls in the search for possible characters which might indicate the cause of death. Why did the two ten-year-old boys die? Their skulls do not reveal any specific features, only that skull 2 has an occipital bone thickened by rickets. Investigations were made to check whether, either in 1372 or in 1444, there had been a rampant epidemic of the plague in Celje. No confirmative response could be found. Had the boys died as a result of child diseases, scarlet fever, diphtheria, smallpox, or from injuries?

Three of the adult counts, Hermann III of Celje, Ulrich II of Celje and Wilhelm, died an unnatural death. Hermann III of Celje, aged between 35 and 45, fell from a horse; Ulrich II of Celje was the victim of a political assassination; as for Wilhelm, the sources claim that in 1392 he returned from a military campaign against the Turks, fell ill in Vienna and died there. He was only 30 years old. There are no data on whether at that time the plague was raging, or spotted fever, or whether perhaps he died from some infection, etc.

Ulrich I of Celje was frequently engaged in combat. The sources record his campaigning in foreign countries. He died aged only 35 to 40, of an unknown cause. Did he, too, die an unnatural death?

The Countess Elisabeth of Frangepan was murdered; she was purportedly killed by her own husband Friedrich II of Celje, in order that he might marry Veronika of Desenice.

One has to ask why the legitimate and illegitimate sons of Hermann II of Celje, Ludwig and Hermann, died »young« as the sources record, at around the age of 30, and why the two countesses Adelaide of Ortenburg and Anna of Schaunberg died at much the same age. With the women-mothers one may also reflect on the bloodtax of their mission. Did Adelaide bear only the son Wilhelm? Among all the Countesses of Celje, Anna of Schaunberg was the one who bore the most children: in historical memory 6 are recorded, the last being Barbara of Celje. Did Anna perhaps die during her following childbirth, or simply from physical exhaustion, or from some disease?

Paleopathological studies reveal that mankind, throughout all periods of history, has been threatened by tuberculosis. The same researches indicate that this specific infectious disease does not leave traces on the skull. One might conjecture that those examined had contracted tuberculosis. If more material was available and if signs of TB were to be found, one could attempt by paleomicrobiological and paleoserological methods to establish the possible presence of the causative factors or of their antibodies. In numerous instances it would be possible to further build up the research with additional paleopathological investigations, if we did not have only the skulls, which are of course also historically protected.

Two of the Celeian couples, Friedrich I and Didmund of Walsee, and Hermann I of Celje and Catherine Kotromanić experienced a long life and a natural death. Among all the Celeians, the greatest age was attained by Friedrich II of Celje – around 80 years.

Among the contagious diseases which usually accompanied the armies, after the discovery of America, the one which imprinted itself as an ugly memory in the soldiers was syphilis. The travelling counts are not expected to have it, since the disease was probably brought to Europe only after the return of Columbus' sailors.

While considering the contraction of diseases, one must reflect also on birth and its opposite, death. Why did such a powerful family die out? At the time of the Counts of Celje, the choice of a bride was primarily in accordance with a good dowry. This was then followed by the maternal role. The countly couple had to ensure heirs. Ulrich I of Celje, Hermann III of Celje, and Ulrich II of Celje remained without descendants. It is not known whether the couples had several children, as is recorded in the genealogy, yet these died young, or whether they actually had so few children. Friedrich I had two sons and two daughters, Ulrich I of Celje only one son, Wilhelm, and Hermann I of Celje had two sons, Hans and Hermann II of Celje. The latter had most children – three sons and three daughters: the son Ludwig died aged 30, without children, Hermann III of Celje also had no children, while Friedrich II of Celje had only the son Ulrich II of Celje. Three children were born to him, but all died early. Infant mortality had been a major problem right up till the 20th c., when the introduction first of vaccination, then of antibiotics and chemotherapeutics, and the improvements in hygienic and social conditions gave protection to children's lives.

One's thoughts also turn involuntarily to the counts' wives, with whom they were united by an agreement between the parents shortly after birth. They were drawn from diverse cultural, language and religious environments, hence the marriages represented an alliance of political and economic power. It would, however, probably have been difficult to prepare the hearts of these spouses to accept their partners – more or less attractive to the eye – and also happily to share with them the wedded state. In short, the male line of the Counts of Celje died out because they did not take enough care for their descendants.

5.3.
Assessment of certain identification methods

5.3.1. X-ray Method

In this research, the radiographs were of crucial importance. High quality X-rays, produced in six projections and in standardized conditions, made it possible to present with the utmost precision all the potential X-ray information, in particular the shape and size of the facial cavities. Damage to individual skulls deprived us of certain data. All radiographs were thoroughly examined, and from them conclusions were also drawn regarding: gender and age, particular pathological changes, ortho- or dysgnathic, and the characteristic shaping features of individual skulls and faces.

Indirect assessment of the age of the skulls from the X-rays, and of the bone thickness and obliteration of the cranial sutures, also proved to be useful. Only for the two child skulls was it possible also to trace the bone nucleus. For a more precise examination of the changes on the mastoids it would be necessary to use additional radiographs.

In certain instances, the X-rays revealed round shadows (lesions), so it was necessary to establish differential diagnosis between potential physiological structures or infected processes, and cystic formations or metastases. In two cases, the answer was provided only by computer tomography.

With the child skulls, the intention was to perform a precise radiograph of the dentition. It was shown that orthopan tomography could be used only for imaging the living, in whom there also remains soft tissue beside the bone tissue. The child dentition, therefore, was studied from classical X-rays. The wisdom teeth were also of interest in all images. Although their outgrowth is determined within a relatively broad time-span (ages 16–30), they did assist in estimating age.

5.3.2. Epigenetic Methods

DNA analysis has provided modern science with the ideal identification method. It is only if DNA cannot be used to provide answers as to identity that we seek other methods. For identification only by skull, for persons buried together and supposedly linked by kinship, the epigenetic method has been developed; in the present research, this was the method of choice and the only one which could lead to scientifically grounded identification.

Regarding the epigenetic relationship between parents and children, conclusions cannot be drawn if one does not have the posthumous remains of both parents; even if these remains are available, one must remain aware that inheritance is polygenic, that in children of the same parents various combinations of genes may occur, expressions of different genes, and consequently the hyperostotic or hypoostotic features, on the child skulls are different. Comparison of the facial cavities is one of the best researched approaches in epigenetic identification. This is why it formed the basis of the research. These cavities were compared between the closest relatives, parents and children.

Comparison of the facial cavities among all the male skulls reveals shared traits. Most of the frontal sinuses are fan shaped, most of the maxillary are triangular, all the orbits are between square and trapezoid shape, and most of the nasal cavities are similar to scallop shape with a serrated margin. The nasal cavities are also similar in size, except that over the centuries of family interment they show a moderate tendency towards shrinking. Such a degree of uniformity in the shape and size of the facial cavities among the male members of the family is certainly no coincidence: it speaks of blood relationship.

By contrast, comparison of the facial cavities by shape and size among the female members of the family is somewhat different. The greater diversity in shape and size is notable. The women did, in fact, come from different environments and brought with them varied hereditary material, and therefore the epigenetic properties of their skulls are different than in blood relatives. This provides a further argument towards the conclusion of this research.

Since the women who married into the family were mostly of relatively small stature, the size of the male members of the family had also somewhat diminished over the centuries.

Determining kinship in a qualitative way is surely closer to the anthropological than to the biomedical discipline. By means of planimetry, an endeavour was made to objectivize the results from the comparison of the facial cavities and to quantify them. The smallness of the sample did not permit statistical verification.

This research was highly specific, interdisciplinary, and did not have as its basis the classical laboratory experiments so typical of biomedicine. Also, consultation with biomedical informants led to the conclusion that, with the 18 skulls which in the study had been grouped into small families, it would simply not be possible from three or four results to calculate the statistical averages, trends and others. This is a task which actually lies between humanistic and natural science research studies.

The new and referential study of this kind on the applicability of the epigenetic method in identifying seven skeletons from a family tomb in Stillfried, Austria was taken as a starting-point. The experts identified them as three adults, sister and brother with wife, and four children, of whom three were from that marriage, while the fourth child was from relations between the brother and sister. The whole family had died simultaneously on account of poisoning and had been buried in a common grave. The two authors of the study, Szilvássy and Hauser, are the most important researchers in the field of epigenetics. Both also acted as consultants for the present study.

5.3.3. Genetic Methods

The incentive for undertaking genetic research was the discovery of living descendants from the line of Barbara of Celje, daughter of Hermann II of Celje. The theoretical expectations concerning the sequences of mitochondrial DNA for this research, which were prepared in the reference genetic laboratory of Dr. Gill – who together with colleagues identified the posthumous remains of the Russian imperial family of the Romanovs – proved in our case to be unfeasible. The posthumous remains of the Romanovs date back to the year 1917, while the Celeian remains are half a millennium older.

In the paleogenetic laboratory it was shown that the possibilities for isolating inherited substances from these old bones were nil, on account of the high state of degradation of the isolated samples and their contamination with human DNA, and also because of the lack of material; therefore, genetic research was not feasible in this instance. It is, however, interesting that in the research on the Romanovs, comparative samples of DNA were obtained from living descendants of the line of Maria Louisa, princess of Hessen and Kassel, who is at the same time of the 16th generation in the line of Barbara of Celje. Clearly, the numerous European nobles are inter-related in closer or more distant kinship, and hence in all likelihood common sequences of those parts of the genomes which are highly polymorphic in humans would appear.

5.4. Questions of Identity

In the central part of the genealogical tree, the research would have gained further credibility had it been possible to locate the skull of Hermann II of Celje. In the family vault one would, in fact, have expected to find the skulls of his three sons from the marriage to Anna of Schaunberg, and hypothetically also of his illegitimate, later legitimized son Hermann. However, all the endeavours of the search for his posthumous remains at the local cemetery of Pleterje proved to be in vain.

Likewise missing from the end of the genealogical record are the posthumous remains of Catherine Branković, which would have influenced the identification of her son Hermann IV of Celje, although this child skull 16, which had remained in the developmental phase, still did not bear all the decisive epigenetic traits and the ultimate shape and size of the facial cavities.

The point at which the identification results deviate from the numbering sequence is with skulls 6 and 7. These belong to the brothers Hermann I of Celje and Ulrich I of Celje. The latter died in 1368, aged approximately 35 according to the historical sources; Hermann I of Celje died as the last adult male in the family, in the year 1385, at the age of over 50. The identification of the skull indicates a younger male skull, in which the facial cavities and other epigenetic traits are similar to skull 8, that of his son Wilhelm. In skull 6, however, we have the skull of a male, aged over 50, whose facial cavities and epigenetic traits are similar to those of skull 2, his son Hans. Since, for these two skulls, the identificational, historical and genealogical data are in accordance, yet the numbers are interchanged, one may conclude that this was due to a mistake in the numbering, which was surely unwittingly made by Anton Zupančič. This conclusion is further supported by many verifiable and reliable professional arguments.

Most of the questions of identity arise about skulls 13 and 14. If they relate to the two illegitimate, later legitimized sons of Hermann II of Celje and Friedrich II, one would expect to find in the vault, beside the former, his halfbrothers Ludwig, Hermann III of Celje and Friedrich II of Celje, and by the latter Ulrich II of Celje. The half brothers were also, in principle, similar to each other; there are also examples, however, of how the polygenicity of inheritance influences the differences. With regard to the three hypothetical brothers and their epigenetic identification, search was made for a female skull which, by the chosen methodology, would suit the traits sought. Skull 10, which conforms not only to the identificational, but also to the historical and genealogical criteria and to the numbering, was compared to the hypothetical sons; this led to the finding that some had inherited more, others less, from her. The relative similarity between skulls 9, 13, 15 and 17, and their epigenetic identification, allows the possibility that among them may also be the illegitimate son Hermann II of Celje. Skull 13 is attributed to him. The question which justifiably arises here, and which is surely not merely rhetorical, is linked to his bishop's status. Could a high-ranking church dignitary have been interred in a family (private) vault? His supposed burial in St Daniel's church in Celje, in the immediate neighbourhood of St Mary's Church, is symbolised by a burial plaque. It is affixed right by the entrance door, yet beside it there is no grave. This would indicate that the plaque had been brought from somewhere else. Had it been at the no longer existing cemetery beside the church? Was Ludwig interred there, and disinterred? And if he was, where are his posthumous remains, and what is to be made of the identity of skull 13? Enquiries at the church and at the deaconate of Celje did not provide any answers. Does the conjecture that the powerful Hermann II of Celje wished to have all his sons – particularly those who did not have a long life – buried in one place, stand up to criticism?

Also in the case of the expected family burial of Hans, illegitimate son of Friedrich (II) in the communal vault, similarities were sought with the father (skull 17) and half-brother Ulrich II of Celje (skull 18) by means of epigenetic identification. Skull 14 stands out from all the others, and in the facial sinuses is very similar to skull 17, that of the hypothetical father, except that the surfaces of the facial cavities are much smaller. Skull 18, the hypothetical half-brother, has somewhat different features to those sought, as does skull 10, supposedly the grandmother. Skull 14 has prominently large fan-shaped-pyramidal frontal sinuses, which are not observed in the other skulls of this research. If this skull is considered as belonging to the illegitimate son of Friedrich II of Celje, then he inherited the facial cavities and most of the epigenetic traits through the mother. If, however, skull 14 was actually to belong to Hans, then it is not suitably numbered regarding the other skulls and the year of death, i.e. the order of interment. The sources record that Hans was legitimized in 1447; this is the only fact altogether that can be traced about him. If he had died shortly thereafter, let us say before the death of Hermann IV of Celje in 1452 – who was the one who died closest to the year 1447 – then Hans's skull should be numbered 15 and not 14. It is mentioned several times that Hermann III of Celje died in 1426 or 1428. Is it possible that Anton Zupančič was mistaken also in this instance, and exchanged the numbers with skulls 14 and 15? The diagnosis of Albright's syndrome with acromegaly is the answer to the special features of skull 14.

5.5.
Commentary to the Artist's Images

In creating the portraits of Friedrich I, Friedrich II of Celje, and Ulrich II, the excellent portrait artist and Academy painter Rudi Španzel has used all the anthropological and anthropometric data which it was possible to offer him. The features on the portraits are most appropriately conveyed: the shape and height of the upper part of the face, the breadth of the face (bizygomatic breadth), height and inclination of the forehead, prominence of the supraorbital arches and glabella, breadth of the nasal root and shape of the nose (but not the length), positioning of the maxilla towards the face (orthognathically or prognathically), prominence of the muscle attachments of the face and skull, which are indicative of a more or less muscular face, and several other fine details. Regarding depiction of the soft tissue, the painter has studied the appropriate literature and has informed himself about the historical role and importance of these three counts, so that he was able to combine their images with the spirit of the times. The remaining original features in the portraits are the result of the artist's creative imagination and experiencing of the historical personalities.

6.
CONCLUSION

The research may be condensed into the following conclusions:

1. The central hypothesis of the research has been proved, i.e. that the skulls are interrelated by kinship and belong to the family of the Counts of Celje, from Friedrich I to Ulrich II of Celje.

2. The hypothesis that the principle of numbering the skulls does not follow the chronology of the burial of all family members has been confirmed. It was established that the males were buried separately from the females and children, and that in doing so the order of the time of death was respected, both for the male and female members alike. There is an error in the numbering of skulls 6 and 7: person 7 died (1368) before person 6 (1385).

3. With an understanding of the principle of numbering the skulls, it was possible to reconstruct the destroyed burial vault and the former disposition of the coffins within it: the deceased were interred in a vault approximately 15 m long and 10 m wide, in four rows; in the first, three women and one child; in the second, five men; in the third, three women; and in the fourth, five men and one child (see point 5).

4. Identification of the skulls by gender:
   - 10 adult males
   - 6 adult females
   - 2 children

5. The identificational, historical and genealogical results, together with the numbering of the skulls, led with a great degree of probability to the following names:

skull 1: Didmund of Walsee (died 1353 or 1354)
skull 2: Hans (died 1372)
skull 3: Adelaide of Ortenburg (died 1391)
skull 4: Catherine Kotromanić (died before 1408)
skull 5: Friedrich I (died 1359 or 1360)
skull 6: Hermann I of Celje (died 1385)
skull 7: Ulrich I of Celje (died 1368)
skull 8: Wilhelm (died 1392)
skull 9: Ludwig (died 1417) (probably, but there is no skull of the father)
skull 10: Anna of Schaunberg (died before 1396)
skull 11: Elisabeth of Frangepan (died 1422 or 1423)
skull 12: Elisabeth of Abensberg (died before 1424)
skull 13: Hermann, legitimized son of Hermann II of Celje (died 1421) – supposedly
skull 14: Hans, legitimized son of Friedrich II of Celje (after 1447) – supposedly
skull 15: Hermann III of Celje (died 1428) (probably, but there is no skull of the father)
skull 16: Hermann IV of Celje (died 1444) (probably, but there is no skull of the mother)
skull 17: Friedrich II of Celje (died 1454) (probably, but there is no skull of the father)
skull 18: Ulrich II of Celje (died 1456)

6. Distribution into families

5 – father Friedrich I
1 – mother Didmund of Walsee,
6 – son Hermann I of Celje,
7 – son Ulrich I of Celje,

7 – father Ulrich I of Celje,
3 – mother Adelaide of Ortenburg,
8 – son Wilhelm,

6 – father Hermann I of Celje,
4 – mother Catherine Kotromanić,
2 – son Hans,

10 – mother Anna of Schaunberg,
9 – son Ludwig,
15 – son Hermann III of Celje,
17 – son Friedrich II of Celje
13 – their half-brother Hermann, Bishop of Freising – supposedly,

17 – father Friedrich II of Celje,
11 – mother Elisabeth of Frangepan,
18 – son Ulrich II of Celje,
14 – his half-brother Hans – supposedly,

15 – husband Hermann III of Celje,
12 – wife Elisabeth of Abensberg,

18 – father Ulrich II of Celje,
16 – son Hermann IV of Celje.

7. Forensic medicine expertise has confirmed that the person with skull 18 succumbed to the consequences of grave head injuries, caused by a sword blow.

List of pictures and illustrations

Illustrations for chapter Methodology of Work

Illustrations for chapter Identification Results

Illustrations for chapter Identification Results: Determining Kinship of Skulls

7.
LITERATURE

7.1.
Natural Science Literature

Anderson DL, Thompson GW. Interrelationships and sex difference of dental and skeletal measurements. J Dent Res 1973; 52: 431–8.

Ascádi GY, Nemeskéri J. History of human life span and mortality. Budapest: Akadémiai Kiadó, 1970: 115–6.

Aufderheide AC, Rodrígez-Martín C. Hematological disorders. In: Human Paleopathology. Cambridge University Press, 1998: 345–351.

Aufderheide AC, Rodrígez-Martín C. The Cambridge Encyclopedia of Human Paleopathology. Cambridge University Press, 1998: 1088–1095, 1158–1169, 1173–82.

Berg S, Rolle R, Seemann H. Der Archäologe und der Tod. Archäologie und Gerichtsmedizin. München-Luzern: Verlag C. J. Bucher, 1981: 145–153.

Bergman P. The occurrence of selected non-metrical traits of the skull in relation to cribra orbitalia and grave equipment. Variability and Evolution, 1993; 2–3: 63–75.

Berry AC, Berry RJ. Epigenetic variation in the human cranium. J Anat 1967; 01: 361–379.

Bikar I. Osnovi ortopedije vilica (Fundaments of orthopaedics of jaws). Beograd: Srpsko lekarsko društvo: 1970: 30–36.

Boddington A, Garland AN, Janaway RC. Dead, decay and reconstruction. Manchester University Press, 1987: 224–239.

Brothwell DR. Ancient Disease. In: Digging up Bones. London: British Museum of Natural History, 1972: 132–170.

Brothwell DR. Cranial Sutures. In: Digging up Bones. London: British Museum of Natural History, 1972: 38–40.

Dominok GW. Zur Alters- und Geschlechtsbestimmung aus der Morphologie der menschlichen Schädelkalotte. Zbl. allg. Path. 1959; 100: 54–63.

Eicgkstedt E. Eine Ergänzung der Weichteile auf Schädel- und Oberkörperskelett eines Neanderthalers. Z. Anat. Entwickl.-Gesch. 1925; 27: 363–80.

Felgenhauer, Szilvássy J, Kritscher H, Hauser G. Stillfried. Archeology – Antropology. Stillfried: Veröffentlichungen des Museums für Ur- und Frühgeschichte Stillfried. Sonderband 3, 1988.

Gerasimov MM. Gesichtsrekonstructionen historischer Persönlicheiten. In: Wissenschaft und Menschheit. Leipzig, Jena, Berlin: Urania Verlag, 1967: 81–105.

Gill P, Ivanov PL, Kimpton C, Piercy R, Benson N. et al. Identification of the remains of the Romanov family by DNA analysis. Nat Genet 1994; 6: 130–135.

Grashey – Birkner. Atlas Typischer Röntgenbilder vom normalen Menschen. München: Urban & Schwarzenberg, 1964:109–182, 718–734.

Grmek MD. Bolesti u osvit zapadne civilizacije (Diseases in the light of the Western Civilization). Zagreb: Globus, 1989: 57–145.

Gustafson G. Forensic Odontology. London: Staphless Press, 1966.

Hagelberg E. Mitochondrial DNA from Ancient Bones. In: Herrmann B, Hummel S. Ancient DNA. Recovery and Analysis of Genetic Material from Paleontological, Archeological, Museum, Medical, and Forensic Specimens. New York-Berlin: Springer, 1992: 195–204.

Harradine N, Suominen R, Stephens C, Brown I. Holograms as substitutes for orthodontic study casts: A pilot clinical trial. Am J Orthod Dentofac Orthop Avg. 1990: 110–116.

Hauser G, De Stefano GF. Epigenetic Variants of the Human Skull. Stuttgart: Nägele & Obermiller: 1989: 2–252.

Hauser G. Epigenetische Merkmale am Gesicht bei den sieben Stillfrieder Skelleten. In: Felgenhauer, Szilvássy J, Kritscher H, Hauser G. Stillfried. Archeology – Antropology. Stillfried: Veröffentlichungen des Museums für Ur- und Frühgeschichte Stillfried. Sonderband 3, 1988: 44–55.

Helwin H. Problematik der Weichteilrekonstruction auf dem menschlichen Schädel. Gegenbaurs morph. Jb. Leipzig 1971; 116: 503–13.

Hengen O.P. Cribra orbitalia: Pathogenesis and probable etiology. Homo, 1971; 22 (2): 57–76.

Herrmann B, Hummel S. Ancient DNA. Recovery and Analysis of Genetic Material from Paleontological, Archeological, Museum, Medical, and Forensic Specimens. New York-Berlin: Springer, 1992: 1–195.

Holland MM, Fischer DL, Mitchell LG, Rodriguez WC, Canik JJ, Merril CR, Weedn VW. Mitochondrial DNA sequence analysis of human skeletal remains: identification of remains from Vietnam war. J Forensic Sci 1993; 38: 542–553.

Hopwood AJ, Mannucci A, Sullivan KM. DNA typing from human faeces. Int J Legal Med 1996; 108: 237–243.

Hrdlička A. Practical Anthropometry. Philadelphia, 1939.

Hunger H, Leopold D, Wunderlich G. Identifikation durch Röntgenaufnahmen. In: Identifikation. Berlin, Heidelberg, New York: Springer Verlag, 1978: 242–261.

Hunger H, Leopold D. Identifikation. Berlin, Heidelberg, New York: Springer Verlag, 1978: 113–133, 162–168, 263–286,

Hunger H, Rother P. Altersbestimmung am Skelett. In: Hunger H. & Leopold D. (Hrsg) Identifikation. Berlin, Heidelberg, New York: Springer Verlag, 1978: 162–182.

Jeffreys AJ, Allen MJ, Hagelberg E, Sonnberg A. Identification of the skeletal remains of Josef Mengele by DNA analysis. Forensic Sci Int 1992; 56: 65–76.

Katolyi L. Paleopathologie (Aufgaben, Objekte, Metodik, gegenwärtige Stand). Sudhoffs Archiv 1970; 54: 398–422.

Kaučič M. Forensic-medical research into material from the vault at the monastery Stična. Prešeren research paper at the Medical Faculty (MF), University of Ljubljana. Mentor Prof. Dr. Anton Dolenc. Ljubljana: MF, 1976.

Keen JA. A study of the differences between male and female skulls. Am J Phys Antropol 1950; 8: 65–79.

Kemkes-Grottenthaler A. Sterbealterbestimmung anhand des ektokranialen Nahtverschlusses: Eine Valuierung der Meindl-Lovejoy-Methode. Rechtsmedizin 1996; 6: 177–84.

Key CA, Aiello LC, Molleson T. Cranial Suture Closure and Its Implications for Age Estimation. International Journal of Osteoarcheology, 1994; 4: 193–207.

Krogman WM, Iscan MY. The human skeleton in forensic medicine. Springfield: Thomas, 1986.

Krogman WM. A problem in the aging of human skeletal remains. J. Forensic Sci 1962; 7: 255–64.

Leopold D. Personnenerkennung durch Superprojektion. In: Hunger H, Leopold D. Identifikation. Berlin, Heidelberg, New York: Springer Verlag, 1978: 263–282.

Loepp W, Lorenz R. Röntgendiagnostik des Schädels. Stuttgart: Thieme; 1954.

Lovrinčević A, Mikić Ž. Atlas osteopatoloških promjena na istorijskim populacijama Jugoslavije (Atlas of osteopathologic changes of historical Yugoslav population). Sarajevo: Svjetlost, 1989.

Lunth LL, Lunth P. Handbook for Dental Identification. Philadelphia-Toronto: Lippincott Comp., 1973.

Lutz S, Weisser HJ, Heizmann J, Pollak S. MtDNA as a tool for identification of human remains – Identification using mtDNA. Int J Legal Med 1996; 109: 205–209.

Manzi G, Vienna A. Cranial non-metric traits as indicators of hypostosis or hyperostosis. Rivista di Antropologia (Roma) 1997; 75: 41–61.

Martin R, Saller K. Lehrbuch der Anthropologie in systematischen Darstellungen der Anthropologischen Methoden. Band I. Stuttgart: Gustav Fischer Verlag, 1957: 362–518; Band II., 1959: 1402–1417.

Milčinski J. Identification. In: Sodna medicina (Forensic Medicine). Secretariat of Home Affairs IRS. Ljubljana, 1958: 39–62.

Møller-Christensen V. Bone changes in leprosy. Copenhagen 1961 (9 f, 16 ft).

Møller-Christensen V. Ten lepers from Naestved in Denmark. A study of skeletons from a medieval Danish hospital. Copenhagen. 1953 (119 f).

Moodie RL. Paleopathology, an introduction to the study of disease. Urbana: University of Illinois Press, 1923.

Moseley JE. Skull Changes in Chronic Iron Deficiency Anemia. American Journal of Roentgenology, Radium Therapy, and Nuclear Medicine 1961; 85: 649–652.

Ortner D, Putschar GJ. Identification of Pathological Conditions in Human Skeletal Remains. Washington: Smithsonian Institution Press, 1985: 29–51, 92–94, 119–144, 162–169, 226–234, 235–139, 270–285, 291–298, 419–430, 436–456.

Ossenberg NS. The influence of artificial cranial deformation on discontinuous morphological traits. Am J Phys Antropol 1970; 33: 357–372.

Perizonius WRK. Closing and non-closing sutures in 256 crania of known age and sex from Amsterdam (AD 1883–1909). J Hum Evol 1984; 13: 201–216.

Powers R. The disparity between known age and age as estimated by cranial suture closure. Man, 1962; 83, 84: 52–55.

Prag J, Neave R. Making faces. Using forensic and archeological evidence. London: British Museum Press, 1989: 12–84.

Puscher E, Szilvássy J. Überlegungen zur grafischen Weichteilrekonstruction nach dem Schädel. Antropologie 1996; 34: 329–339.

Rämsch R, Herrmann WR. Ergebnisse von Messungen der Schädelkalottendicke bei Männern und Frauen. Dtsch. Z. ges. gerichtl. Med 1963/64; 54: 227–30.

Sassouni V. Palatoprint, physioprint, and roentgenographic cephalometry, as new methods in human identification. J Forensic Sci 1957; 2: 428–442.

Schutkowski H, Grupe G. Zusammenhänge zwischen Cribra orbitalia, archäometrischen Befunden am Skelett und Habitatbedingungen. Antropologische Anzeiger 1997; 55: 155–166.

Schwidetzky I. Bemerkungen zur Geschlechtsdiagnose am Schädel. Wiss. Z. Humboldts-Univ. Berlin. Math.-Naturwiss. Reihe 1969; 18: 965–7.

Smodlaka V. Antropometrijska tehnika (Anthropometric technique). Beograd: Medicinska knjiga, 1948.

Stadtmüller F. Die Rekonstruktion der Weichteile des Schädels. Mat. Medica Nordmark 1961; 5: Sonderheft.

Steinbock RT. Paleopathological diagnosis and interpretation. Springfield-Illinois: Charles C. Thomas, 1976.

Stewart TD. Sex determination of the skeleton by guess and by measurements. Am J. Phys Antropol 1954; 12: 385–92.

Sullivan KM, Mannucci A, Kimpton CP, Gill P. A rapid and quantitative DNA sex test: fluorescence-based PCR analysis of X-Y homologous gene amelogenin. Bio Technique 1993; 15: 636–641.

Szilvássy J, Kritscher H, Vlček E. Die Bedeutung röntgenologischer Methoden für die antropologische Untersuchung ur- und frühgeschichtlicher Gräberfelder. Ann. Naturhist. Mus. Wien 1987; 89: 313–352.

Szilvássy J. Die Stirnhöhlen der Schädel aus dem awarischen Gräberfeld von Zwölfaxing in Niederösterreich. Ann. Naturhistor. Mus. Wien 1974; 78: 109–118.

Szilvássy J. Eine neue Methode zur intraserialen Analyse von Gräberfeldern. In: Innovative Trends in der prähistorischen Antropologie. Ed. Herrmann B. Göttingen, 1986: 51–62.

Szilvássy J. Zur Entwicklung der Stirnhöhlen. Anthrop. Anz. 1981; 39: 138–149.

Szilvássy J. Zur Variation, Entwicklung und Vererbung der Stirnhölen. Ann. Naturhist. Mus. Wien 1982; 84A: 97–125.

Škerlj B. Anthropometry and anthroposcopy. In: Medical Encyclopaedia. Zagreb: Jugoslavenski leksikografski zavod, 1957, vol–1: 484–488.

Španzel R. The Counts of Celje in the eyes of Rudi Španzel. Celje: Regional Museum; Ljubljana: Geodetic Institute of Slovenia, 1999: graphic map in metal covers with six graphics.

Tabor L, Jevtić V, Pavčnik D. Rentgenologija (Roentgenology). Ljubljana: Medicinski razgledi (Medical Views), 1990.

Tomazo Ravnik T, Ivanović B. Anthropologic characteristics of historical personages of the Petrović-Njegoš Dynasty. Etnoarheološki problemi (Beograd) 1992; 9: 121–138.

Trimble MK. Landmarks of the Skull Used for Taking Measurements and Calculating Indices. In: Human Osteology – A Laboratory and Field Manual. Columbia: Missouri Archeological Society, 1987: 62–79.

Vlček E, Boubín J. Hold českému králi Jiřimu z Podebrad. Praze: Národní muzeum v Praze, 1991: 2–24.

Zuhrt R, Rottstock F, Winterfeld RI. Möglichkeiten und Methoden der Stomatologie bei der Identifizirung. In: Hunger H, Leopold D. Identifikation. Berlin, Heidelberg, New York: Springer Verlag, 1978: 287–310.

Yoshino M, Matsuda H. et al Computer-assisted identification system using video superimposition. Forensic Sci Int1997; 90: 231–122.

7.1.1. Additional Natural Science Literature

Berg S, Rolle R, Seemann H. Auf der Suche nach Personen und Gesichtern. In: Der Archäologe und der Tod. München-Luzern: Verlag C.J. Bucher, 1981: 145–152.

Antolič I, Demirović D, Farčnik F, Lapter V, Ljupči B. et al. Ortodoncija (Orthodonty). Beograd Ljubljana: Mladinska knjiga, 1982.

Ashley KF. Identification of Children in a mass disaster by estimation of dental age. Brit Dent J 1970; 129: 167–73.

Boddington A, Garland AN, Janaway RC. Death, decay and reconstruction. Manchester University Press, 1987.

Cattaneo C, Smillie DM, Gelsthorpe K, Piccinini A, Gelshorpe AR, Sokol RJ. A simple method for extracting DNA from old skeletal material. Forensic Science International 1995; 74: 167–74.

Ferembach D, Schwidetzky I, Stloukal M. Empflehungen für die Alters- und Geschlechtsdiagnose am Skelett. Homo 1979; 30: (I): 32.

Höss M, Handt O, Pääbo S. Recreating the Past by PCR. In: Mullis KB, Ferre F, Gibbs RA. The Polymerase Chain Reaction. Boston Basel Berlin: Birkhäuser, 1994: 257–64.

Hunt EF, Gleiser I. The estimation of age and sex of preadolescent children from bones and teeth. Am J Phys Anthropol 1955; 13: 479–87.

Ivanović B, Tomazo Ravnik T. Skeletni ostaci Lješa i Ivana Crnojevića (Skeletal remains of Lješa and Ivan Crnojević). Titograd: Crnogorska akademija nauka i umjetnosti 1992; 8–9: 119–131.

Kališnik M. (ed.) Pravopis medicinskih izrazov (The orthography of medical expressions). Ljubljana: Medical Faculty, University of Ljubljana, 1997: 611 pp.

Kallay J. Dentalna antropologija (Dental anthropology). Zagreb: Izdavački zavod Jugoslavenske akademije, 1974.

Krušič V. Karies pri starih Slovanih (Caries among the ancient Slavs). Ljubljana: SAZU, 1954.

Lee HC, Ruano G, Pagliaro EM. DNA analysis in human bone and other specimens of forensic interest: PCR typing and testing. J Forensic Sci Soc 1991; 31 (2): 213–6.

Meschan I. An Atlas of Anatomy Basic to Radiology. Philadelphia: W. B. Saunders Company, 1975.

Møller-Christensen V. Osteoarcheology as a medicohistorical auxiliary science. Med Hist 1973; 17: 411–8.

Mullis KB, Ferre F, Gibbs RA. The Polymerase Chain Reaction. Boston Basel Berlin: Birkhäuser, 1994.

Pääbo S. Ancient DNA. Sci Am 1993; 269: 86–92.

Pääbo S. Cloning of Egyptian mummy DNA. Nature (London) 1985; 314: 644–5.

Pepper OHP, Pendergrass EP. Hereditary Occurrence of Enlarged Parietal Foramina. Am J Roentgenol 1936; 35: 1–8.

Rant J. Čeljustna in zobna ortopedija (Maxillary and dental orthopaedics). Ljubljana: Medical Faculty, 1970.

Saunders S, DeVito C, Herring A, Southern R, Hoppa R. Accuracy Tests of Tooth Formation Age Estimation for Human Skeletal Remains. Am J Phys Antropol 1993; 92: 173–188.

Schmitt HP, Saternus K. Beiträge zur forensischen Osteologie. Der Processus mastoideus als Identifikationsmerkmal? Z. Rechtsmedizin 1970; 67: 170–4.

Schumacher GH, Ehler E, Brune K, Pfau H. Untersuchungen über die geometrische Form des Zahnbogens bei Erwachsenen. Dtsch. Stomat. 1970; 20: 729–37.

Steinbock RT. Cribra orbitalia. In: Paleopathological diagnosis and interpretation. Springfield-Illinois: Charles C. Thomas, 1976: 239–252.

Szilvássy J, Kritscher H, Teschler-Nicola M. Zur Verwandschaftanalyse von Skelleten aus ur- und frühgeschichtlichen Gräberfeldern. Ann. Naturhist. Mus. Wien 1987; 88: 183–205.

Štefančič M. Antropološka analiza lobanje z ljubljanskega barja (Anthropological analysis of a skull from the Ljubljansko barje - Ljubljana marshes). Report on paleolithic, neolithic and eneolithic research in Slovenia, 1985; 12: 75–80.

Titiev M. The Science of Man. Nex York, Chicago, San Francisco, Toronto: Holt, Rinehart and Winston, Inc., 1964: 668.

Tomazo Ravnik T. Staroslovansko grobišče v Zgornjem Dupleku (Ancient Slavic burial-site at Zgornji Duplek). Arheoloski vestnik 1979; 30: 489–497.

Vasilić M, Cuculj V, Milošević A. Značaj usta i zuba za identifikaciju (Importance of mouth and teeth for identification). Stomatološki glasnik Srbije 1972; 19: 132–8.

7.2.

Historiographical Literature

Baš F. Celjski grofi in njihova doba (The Counts of Celje and their age). Celjski zbornik, 1951: 7–22.

Baš F. Celjski grofi in njihova doba (The Counts of Celje and their age). In: Contributions to the history of north-eastern Slovenia. Maribor: Obzorja, 1989: 323–35.

Cevc E, Gestrin F, Bregant T, Klaić N, Grafenauer B. et al. Celjani v slovenskem in srednjeevropskem prostoru (The Celeians in Slovene and Central European space). Zgodovinski časopis (Historical Review) 1983; 37 (1-2): 93–114.

Cusin F. Il Confine Orientale d'Italia nella politica del XIV e XV Secolo. II. vol. Milano: A. Giuffré, 1937.

Cusin F. Il Confine Orientale d'Italia nella politica del XIV e XV Secolo. II. Edit. Trieste: Ediz. LINT, 1977: 551 str.

Dopsch H. Die Grafen von Cilli. Ein Forschungsproblem. Süddeutsches Archiv 1974/75; 17/18: 9–49.

Grafenauer B. Celjski grofje (The Counts of Celje). In: Enciklopedija Slovenije (Encyclopaedia of Slovenia), vol. 2. Ljubljana: Mladinska knjiga, 1988: 12–15.

Gruden J. Zgodovina slovenskega naroda (History of the Slovene People): Mohorjeva družba, 1992: 216–69.

Habjan V. Mejniki slovenske zgodovine (Milestones in Slovene History). Ljubljana: Enotnost, 1996.

Habjan V. Vojna za kneževino Celjskih in mir v Požarnici (War for the Principality of the Celeians and Peace in 1460 at Požarnica (Pusarnitz) in Carinthia). Zgodovinski časopis (Historical Review) 48 (1994): 4: 457–498.

Habjan V. Radovljica v času združitve Ortenburških in knezov iz Celja (Radovljica at the time of the union between the Ortenburgs and the Princes of Celje). Radovljiški zbornik 1995: 112–131.

Hartman B. Celjani v slovenskem in srednjeevropskem prostoru (The Celeians in Slovene and Central-European Space). Zgodovinski časopis (Historical Review) 1983; 37: 101–2.

Klaić N. Zadnji knezi Celjski v deželah sv. Krone (The last Celeian princes in the lands of the Holy Crown). Celjski zbornik 1982. Special edition. Celje, 1982: 130 pp.

Kos M. Zgodovina Slovencev (A History of the Slovenes). Ljubljana, 1933: 213–22.

Kos D. Celjska knjiga listin I (The Celje Book of Documents I). Ljubljana-Celje: ZRC SAZU, Muzej novejše zgodovine Celje (Celje Museum of Recent History), 1996: 296 pp.

Kreft B. Celjski grofje (The Counts of Celje). Drama from the life of the medieval feudalists for whom our predecessors lived in bondage. Ljubljana: Tiskovna zadruga, 1932: 130 pp.

Krones F. Die Cillier Chronik. Kritische Untersuchungen Ihres Textes und Gehaltes. Archiv für Österreische Geschichte. Wien, 1837: 1–102.

Krones F. Die zeitgenössichen Quellen zur Geschichte der Grafen von Cilli, mit Einschluss der sogenannten »Ciller Chronik 1341–1456«. Beitrege zur Kunde steiermärkischen Geschichtsquellen. Graz, 1871.

Krones F. Die Freien von Saneck und ihre Chronik als Grafen von Cilli. Graz. 1883.

Krones F. Chronicle of the Counts of Celje. Translation by Golia Maribor: Obzorja, 1972: 7–153.

Marolt M. Cerkev Marijinega vnebovzetja, nekdaj minoritska cerkev (The Church of Mary's Assumption, former Minorite Church). In: Dekanija Celje (Deaconate of Celje). Vol. 1: Church Monuments in Celje. Celje: Zgodovinsko društvo (History Society) in Maribor: 1931: 60–73.

May G. Cilli, Stadt, Landschaft, Geschichte. Cilli: Verlag des Kulturamtes der Stadt Cilli, 1943: 39 strani s 45 slikami.

Mlinarič J. Kartuzija Pleterje (The Carthusian Monastery of Pleterje) 1403–1595. Ljubljana: Kartuzija Pleterje, 1982.

Mlinarič J. Kartuziji Žiče in Jurklošter (The Carthusian Monasteries of Žiče and Jurklošter). Maribor: Obzorja, 1991.

Mlinarič J. Celjani in njihov odnos do samostanov (The Celeians and their relation towards the monasteries). In: Celjski zbornik (ed. Fugger-Germadnik R.). Celje: Pokrajinski muzej (Regional Museum) Celje. 1999: 125–142.

Orožen I. Celska kronika (The Chronicle of Celje). In: Celi 1854, pp. 20–92.

Orožen J. Zgodovina Celja II. del (History of Celje Part II). Medieval Celje as a centre of state-founding aspirations: the era of national independence, the March of Savinja, Princes and Counts of Celje. Celje: Published by Goričar & Leskovšek, 1927: 18–125.

Orožen I. Celska kronika. Faksimile (The Chronicle of Celje. Facsimile). Celje: Osrednja knjižnica, 1997: 25–92.

Otorepec B. Listine grofov Celjskih I. (Registers of the Counts of Celje I). Regesti Haus-, Hof- and Staatsarchiv Wien (Repertory XIV) 1314–1456. Manuscript (100 pp).

Otorepec B. Listine grofov Celjskih II. (Registers of the Counts of Celje II). Regesti Haus-, Hof- and Staatsarchiv Wien (Repertory XIV/1 (Celeian I): pp. 1–25; I (Celeian II: pp. 1–32; XIV/14, Bd.1 (Ortenburg III: pp. 1–10) Manuscript.

Potočnik M. Vojvodina Koroška (Duchy of Carinthia) I. Ljubljana, 1909: 115.

Raukar T. Grofovi Celjski I hrvatsko kasno srednjevekovje (The Counts of Celje and the Croatian Late Middle Ages). Historijski zbornik 1983; 36: 113–140.

Stegenšek A. Konjiška dekanija (Deaconate of Konjice), Maribor, 1909: 179–290.

Štih P. Goriški grofje, (Goerz Counts). Zgodovinski časopis (History Journal) 1994; 48: 499–518.

Štih P. Celjski grofje, vprašanje njihove deželnoknežje oblasti in dežele Celjske (The Counts of Celje, the question of their regional princely authority and of the Celeian lands). Grafenauer's Papers (ed. Rajšp V, Gestrin F.): SAZU, ZRC SAZU, Faculty of Philosophy, Pedagogical Academy Maribor, 1996: 227-256.

Štih P. Celjski grofje – še vedno raziskovalni problem? (The Counts of Celje – still a research problem?) In: Celjski zbornik (ed. Fugger-Grmadnik R.). Celje: Pokrajinski muzej Celje, 1999: 13–22.

Thomas C. Genealogical tree of the Counts of Celje after Dr. Christiane Thomas, 1991.

Voje I. Romanje Ulrika II. Celjskega v Kompostelo k sv. Jakobu (Pilgrimage of Ulrich II of Celje to Compostela St Jacob's). Zgodovinski časopis 38 (Historical Review), 1984; 3: 225–230.

7.2.2. Additional historiographical literature

Bregovac-Pisk M. Portreti grofova Celjskih u sjevernozapadnoj Hrvatskoj (Portraits of the Counts of Celje in northwestern Croatia). In: Celjski zbornik (ed. Fugger-Germadnik R.). Celje: Pokrajinski muzej Celje, 1999: 375–384.

Detela F. Veliki grof. (The Grand Count.) Ljubljana: Karantanija, 2000: 179 pp.

Grdina I, Štih P. Spomini Helene Kottaner: ženski glas iz srednjega veka (Memoirs of Helena Kottaner: a woman's voice from the Middle Ages). Ljubljana: Nova revija, 1999: 124 pp. (Zbirka Korenine / Nova revija).

Grdina I. Celje Princes in Europe. Celje: FIT-MEDIA, 1994.

Habjan V. Friderik II. Celjski – nesojeni kralj Bosne in njegova Veronika iz kajkavskih Desenic (Friderick II. The Celeian not-fated-to be King of Bosnia and his wife Veronika from kajkavske Desenice). Obzornik 1988; 10: 722–729; 11: 787–801; 12: 843–855.

Habjan V. Prvi zapisani in objavljeni ljubezenski stih (The first noted and published Slovene love verses). Naši razgledi, 24 December 1966.

Jurčič J. Veronika Deseniška (Veronika of Desenice). In: Zbrano delo, book nine. Ljubljana: DZS, 1960: 65–123.

Kühn D. Ich Wolkenstein. Frankfurt am Main: Insel Verlag, 1980.

Kmecl M. Troje sodobnih dram (Three Modern Dramas). Zbirka Svet ob Muri. Murska sobota: Pomurska založba, 1965, 283–306.

Kočevar F. Mlinarjev Janez, slovenski junak ali vplemenitba Teharjanov (Mlinar's Janez – Slovene hero or ennoblement of inhabitants of Teharje). Celje: D. Hribar, 1892: 150 pp.

Krajevni leksikon Dravske banovine (Regional Lexicon of the Drava Banovina). Ljubljana, 1937.

Meyer T. Die Grafen von Cilli als Erben der Grafen von Ortenburg. V: Celjski zbornik (ur. Fugger-Germadnik R.). Celje: Pokrajinski muzej Celje, 1999: 85–94.

Neumann HJ. Erbkrankheiten in europäischen Fürstenhäusern. Berlin: Edition q, 1993.

Novačan A. Celjska kronika (dramski mozaik v treh delih). (The Chronicle of Celje (dramatic mosaic in three parts). Herman Celjski (drama v petih dejanjih). (Hermann of Celje (drama in five acts). Ljubljana: Tiskarska zadruga R.Z.Z.O.Z., 1928, 111 pp.

Orožen J. Celje z zaledjem (Celje with its hinterland). Celje, 1948.

Predovnik K. Kartuzija Žiče (The Carthusian Monastery of Žiče). Ljubljana: Department of Archaeology, Faculty of Philosophy, Ljubljana University, 1996.

Rot J. Nagrobni govor o smrti svetlega grofa Ulrika Celjskega (Funeral speech on the death of the illustrious Count Ulrich of Celje). In: Simoniti P. Humanism in the Slovene Lands (and Slovene humanists up till the mid-XVI c.). Ljubljana: Slovenska matica, 1979: 239–244.

Sitar S. Z Desenic Veronika z Desenic (With Veronika of Desenice). Ljubljana: Partizanska knjiga, 1974: 130 pp.

Tringli I. Die Würden und die Besitzungen der Cillier in Ungarn. V: Celjski zbornik (ur. Fugger-Germadnik R.). Celje: Pokrajinski muzej Celje, 1999: 115–124.

Tumograjska J. Nedolžnost in sila (Innocence and Force), 1851.

Vachta B, Pohl W, Vocelka K. The Habsburgs. History of a European Family. Ljubljana: Mladinska knjiga, 1994: 70–76.

Valvasor JV. Slava vojvodine Kranjske (Glory of the Duchy of Carniola). Ljubljana: Mladinska knjiga, 1984, p. 212 on the Celje counts, p. 242 on Hermann II and Pleterje.

Vignjević T. Spomin na Celjske grofe (In Memory of the Counts of Celje). Two portrayals from the late Middle Ages and the Renaissance. In: Celjski zbornik (ed. Fugger-Germadnik R.). Celje: Pokrajinski muzej Celje, 1999: 427–431.

Wambrechtsamer A. Grofje Celjski danes in nikdar več (Counts of Celje Today and Never Again). Maribor: Založba Obzorja, 1977: 540 pp.

Žontar J. Zgodovina mesta Kranja (History of the town of Kranj). Kranj, 1939.

Župančič O. Veronika z Desenic (Veronika of Desenice). A tragedy in five parts. Ljubljana: Zvezna tiskarna in knjigarna, 192k: 185 pp.

Abstract

This original work represents an interdisciplinary identificational research of 18 skulls (without lower jaws), the only posthumous remains, which presumably belonged to the Central-European noble family, the Counts of Celje, who in the 15th century attained power comparable to that of the Habsburgs. Between the years 1350 and 1450 they buried their family members into the family vault in the Minorite church in Celje. When, after a fire, the church was renovated (in 1811), they opened the vault, numbered the skulls by an unknown principle and subsequently stored them. The research, based on most up-to-date approaches, identified those skulls and established their kinship relations. We used the methodology of forensic science, physical anthropology, epigenetics, roentgenology, paleopathology, dentistry and history.

This work is basically divided into three parts: the first part collects and critically evaluates the necessary historical and genealogical data; the second identifies the skulls by gender, age and some other features; the third searches for family relationship between the skulls and compares the historiographic and identification results.

The Counts of Celje still have living descendants on the female side (after Barbara of Celje), who represent the 20th generation. Therefore, the research was directed into genetics. The researchers of the reference laboratory of paleogenetics (G. F. De Stefano, Rome) endeavoured to isolate the mitochondrial DNA from the skull samples. The isolation, however, was not successful.

The identification by following of the epigenetic traits similarity (G. Hauser, Vienna) gave significant results: many morphological features of the skulls are genetically determined, in particular the shape (observed by X-rays) and dimensions (measured by morphomat) of the facial cavities which, in comparison with the frontal sinuses and upper jaw sinuses, as well as orbital and nasal cavities (J. Szilvássy, Vienna), led to the anticipation of blood relation. This was complemented and confirmed by many other examined and compared epigenetic traits on the skulls (approximately 30 per skull). This identification, together with historiography and genealogy, led to determination of the identity of the majority of the skulls and made it possible to group them into families.

By the identification of skulls we also obtained the data on life expectancy and gender, nutritional habits, health and illness (E. Strouhal, Prague), degeneration characters, injuries, individual as well as common face and skull features, dental and orthodontic features of the upper jaw and upper part of the face and many other details.

Based on the identification analyses, anthroposcopic and anthropometric parameters of three male skulls the Academy painter R. Španzel (Ljubljana) also made portraits (Gerasimov's method).

The study brought new comprehension of natural and social sciences and gave interdisciplinary answers. From the sociological, cultural and historical point of view it indirectly reveals the role of men, women and children in that period of time and explains the reasons for separate interment of male (blood) and female (non blood) family members. Even the fact that the skulls presumably belonging to the Counts of Celje are still preserved, points to a deeply rooted attitude of people towards civilisation norms, as preserved in ancient cultures. The dead body – especially when it belonged to a sovereign – was a relic, which should be kept in a safe place. It is not a coincidence that the medieval rulers were buried in churches; it reflects the cultural attitude of those times.

The research could serve as a methodological basis for epigenetic identificational studies of family linked skulls buried in collective graves, especially in cases where genetic identification is not possible (e.g. Bosnia). This work also demonstrates the advantage of interdisciplinary cooperation. The general applicability could be verified in similar identification studies. In addition to its anthropological and medical significance, this research is also of considerable historical interest.

About the Author

– Zvonka Zupanič Slavec was born in 1958 in Slovenia

– Graduated from School of Medicine of Ljubljana University (Slovenia) in 1984

– Master's degree in medicine at the University of Ljubljana in 1991

– PhD in medicine at the University of Ljubljana in 2001

– From 1984 till 1987 worked as a medical doctor at the Institute of Anaesthesiology of Ljubljana Medical Centre

– Since 1987 has been working at the Institute for the History of Medicine at Ljubljana School of Medicine at the University of Ljubljana

– Specialised in History of Medicine at Padua, Vienna, Washington University, and at Wellcome Institute for History of Medicine, London

– Since 1992 has been lecturing on the subjects of History of Medicine and History of Dentistry at Ljubljana School of Medicine at the University of Ljubljana

– Research work is mostly dedicated to the national history of medicine in the 19th and 20th century

– Publishing research books, monographs, research articles from the mentioned fields, mostly in Slovene

– Assist. Professor from 1992, University Professor since 2002

Lek's development has long been based on Slovene scientists and their discoveries. And it will remain this way in the future. We have been developing active pharmaceutical ingredients and pharmaceutical products for 60 years, following modern developments in medicine and pharmacy, and opening doors to the future.

Our goal is to discover and manufacture safe, effective, high-quality and patient-friendly medicinal products. This is what we invest our resources and knowledge in and why we motivate people. Our research successes to date would not have been possible without cooperation with a highly branched network of individuals and institutions in Slovenia. This cooperation is the guarantee of success in the fight against disease, for a healthy future.

lek

a new Sandoz company

Lek Pharmaceuticals d.d. Verovškova 57, 1526 Ljubljana, Slovenia • www.lek.si